Cambridge Primary
Mathematics

Second Edition

Learner's Book 1

Catherine Casey
Steph King
Josh Lury

Series editors:
Paul Broadbent
Mike Askew

Boost

HODDER
EDUCATION
AN HACHETTE UK COMPANY

Cambridge International copyright material in this publication is reproduced under licence and remains the intellectual property of Cambridge Assessment International Education.

Third-party websites and resources referred to in this publication have not been endorsed by Cambridge Assessment International Education.

Registered Cambridge International Schools benefit from high-quality programmes, assessments and a wide range of support so that teachers can effectively deliver Cambridge Primary. Visit www.cambridgeinternational.org/primary to find out more.

The audio files are free to download at www.hoddereducation.com/cambridgeextras.

Acknowledgements

The Publishers would like to thank the following for permission to reproduce copyright material.

Photo credits
p. 46 *cc* © Hachette UK; **p. 113** *cr* © Hachette UK; **p. 130** *br* © C Squared Studios/Getty Images.

t = top, *b* = bottom, *l* = left, *r* = right, *c* = centre

Every effort has been made to trace all copyright holders, but if any have been inadvertently overlooked, the Publishers will be pleased to make the necessary arrangements at the first opportunity.

Hachette UK's policy is to use papers that are natural, renewable and recyclable products and made from wood grown in well-managed forests and other controlled sources. The logging and manufacturing processes are expected to conform to the environmental regulations of the country of origin.

Orders: please contact Hachette UK Distribution, Hely Hutchinson Centre, Milton Road, Didcot, Oxfordshire, OX11 7HH. Telephone: +44 (0)1235 827827. Email education@hachette.co.uk Lines are open from 9 a.m. to 5 p.m., Monday to Friday. You can also order through our website: www.hoddereducation.com

ISBN: 978 1 3983 0090 3

© Catherine Casey, Steph King and Josh Lury 2021

First published in 2017

This edition published in 2021 by

Hodder Education,

An Hachette UK Company

Carmelite House

50 Victoria Embankment

London EC4Y 0DZ

www.hoddereducation.com

Impression number   10 9 8 7 6 5
Year   2025 2024

Cover illustration by Lisa Hunt, The Bright Agency

Illustrations by Alex van Houwelingen, Ammie Miske, James Hearne, Jeanne du Plessis, Natalie and Tamsin Hinrichsen, Steve Evans, Val Myburgh, Vian Oelofsen

Typeset in FS Albert 17/19 by IO Publishing CC

Printed in India

A catalogue record for this title is available from the British Library.

MIX
Paper | Supporting responsible forestry
FSC™ C104740
FSC
www.fsc.org

Contents

How to use this book

This book will help you to learn about mathematics.

Explore the picture or problem.
What do you see?
What can you find?

This icon shows you that the activity links with other subjects in your school curriculum.

Understand new **Maths words**. The *Mathematical dictionary* at the back of this book can help you.

Remember to write any answers in your notebook, not in this textbook.

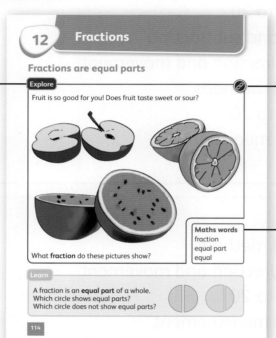

Learn new mathematics skills with your teacher. Look at the pictures to help you.

The shaded questions show you what you need to do.

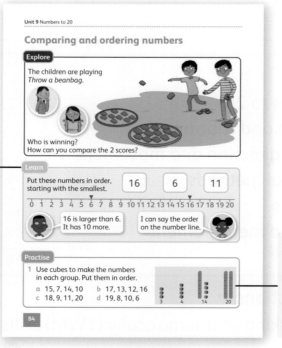

Do the **Practise** activities to learn more. Work like a mathematician.

Try this activities make you think carefully about mathematics.

For **Let's talk** activities, talk about your ideas.

This star shows you the activities that require you to think and work mathematically.

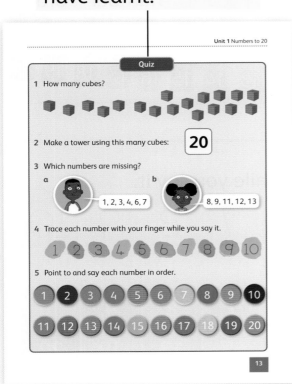

Do each **Quiz** to find out how much you have learnt.

This icon shows you that audio material is available. Listen and you will learn.

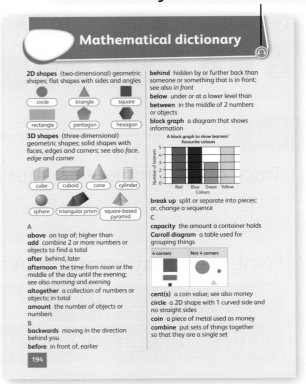

1 Numbers to 20

Reading and writing numbers

Explore

Viti tidied up.

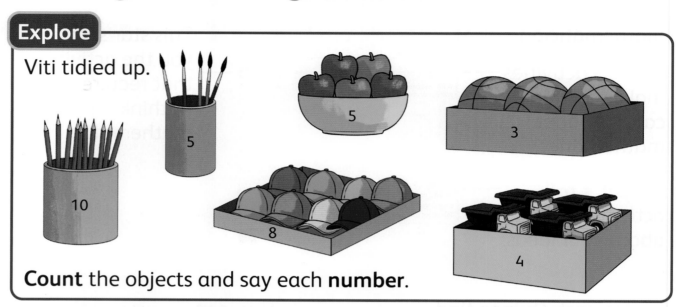

Count the objects and say each **number**.

Learn

Maths words
count   number   order

Say each number from 1 to 20.

1 2 3 4 5 6 7 8 9 10 11 12 13 14 15 16 17 18 19 20

Practise

1 Trace each number with your finger while you say it.

1 2 3 4 5 6 7 8 9 10

2 Point to and say each number in **order**.

1	2	3	4	5	6	7	8	9	10
11	12	13	14	15	16	17	18	19	20

Practise *(continued)*

3 Point to and count along each number track.
Say the missing numbers.

a

1	2	3		5	6		8		10

b

11	12	13			16	17	18		20

c

5		7		9		11		13	

d

1	2		4	5	6	7		9	10

e

11		13		15			18		20

Try this

Count and point.
Count up.
Count down.
Count without looking!

0	1	2	3	4	5
	6	7	8	9	10
	11	12	13	14	15
	16	17	18	19	20

Let's talk

Count up but miss a number. Ask: Which number did I miss?

| 0 | 1 | 2 | 3 | 4 | 5 | 6 | 7 | 8 | 9 | 10 |

Use the number line to help you.

1, 2, 3, 5, 6, 7, 8, 9

I think you missed 4.

Counting objects

Explore

Count the towers.
What **pattern** can you see?

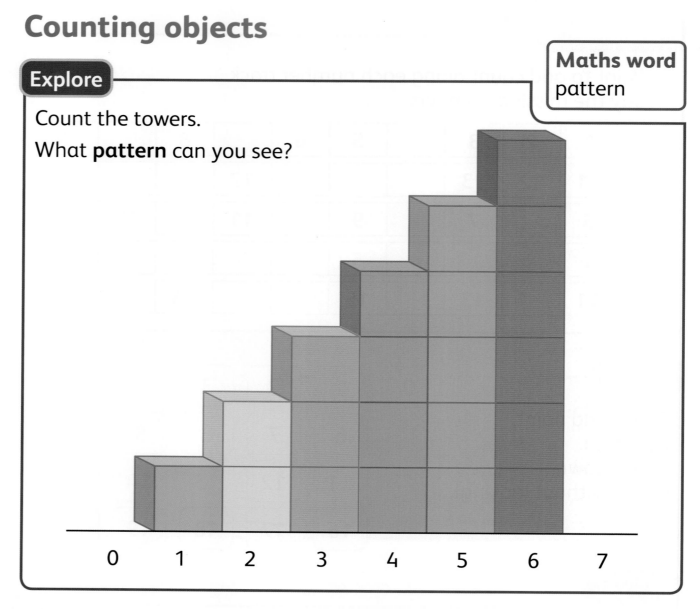

0   1   2   3   4   5   6   7

Learn

How many balls do you see?
Say how many. Then count to check.

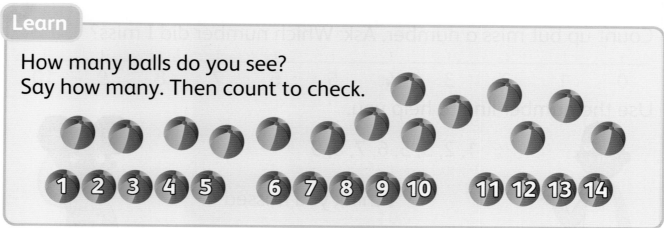

1 2 3 4 5   6 7 8 9 10   11 12 13 14

Practise

1 Make each tower of cubes.

a
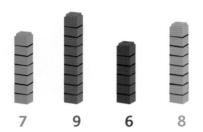

7   9   6   8

b
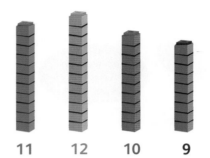

11   12   10   9

2 Count the shapes.

a
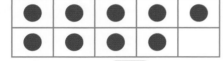

There are ☐ circles.

b

There are ☐ triangles.

c

There are ☐ squares.

d

There are ☐ stars.

3 Count as you do these.

a Draw 8 circles.

b Make a pile of 10 books.

c Draw 17 triangles.

Try this

Count the different fruits.

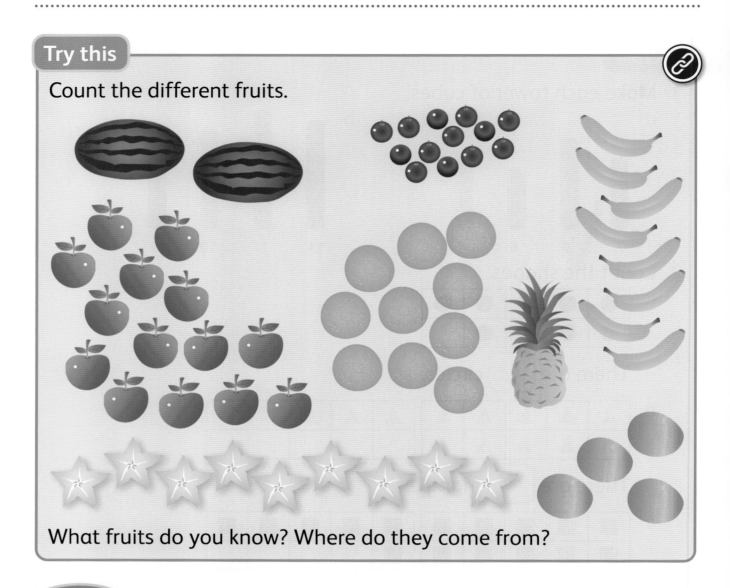

What fruits do you know? Where do they come from?

Let's talk

Work together.
Look away.
Listen and count.
Say how many.
Count and check.
Swap places.

Recognising numbers

Explore

We can show **six** in different ways.

Make **6** from cubes.
Move the cubes around. The number stays the same!

Learn

Do not count. What numbers can you see?

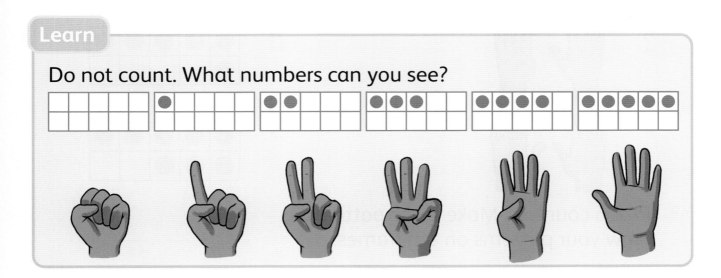

Practise

1   Use cubes or counters to
    make the numbers 1 to 10.
    Use a ten frame like this.

Practise (continued)

2   Match the fingers to the dots.

a   1

b   2

c   3

d   4

e   5

3   Take 6 counters. Make some patterns.
    Draw your patterns on ten frames.

Try this

Work with a partner.

Make a pattern with 3, 4 or 5 cubes.

Hide it behind a book, then show it quickly.

Ask your partner: How many cubes did I use?

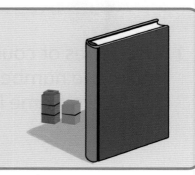

Quiz

1 How many cubes?

2 Make a tower using this many cubes:

3 Which numbers are missing?

a

1, 2, 3, 4, 6, 7

b

8, 9, 11, 12, 13

4 Trace each number with your finger while you say it.

5 Point to and say each number in order.

Addition and subtraction

One more, one less

Look at this tree.
Say what you see.

Maths words
more
more than
less than

Learn

There are 3 monkeys on this number line.

0   1   2   ③   4   5   6

1 **more** monkey climbs on.

0   1   2   ③   ④   5   6

There are now 4 monkeys. We say: 1 **more than** 3 is 4.

There are 6 parrots on this number line.

0   1   2   3   4   5   ⑥

1 parrot flies away.

0   1   2   3   4   ⑤   ⑥

There are 5 parrots left. We say: 1 **less than** 6 is 5.

Practise

Maths word
less

1 Look at the shapes on this number line.

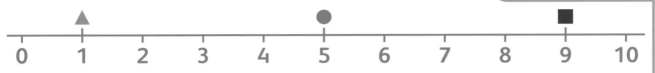

a Which number is 1 less than the ●?

b Which number is 1 less than the ■?

c Which number is 1 more than the ▲?

d Which shape is 1 less than 6?

2 Find the missing numbers to show 1 more or 1 **less**.

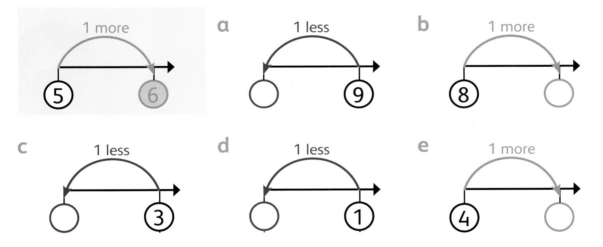

Try this

David thinks of a number but says the number that is 1 less.

Maris thinks of a number but says the number that is 1 more.

Both David and Maris say the **same** number.

What number are they both thinking of?

Tell a partner how you know.

Counting on and counting back on a number line

Maths words
add
subtract
take away

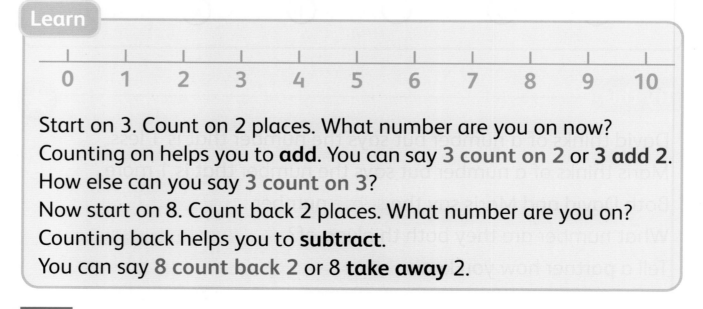

Start on 3. Count on 2 places. What number are you on now?
Counting on helps you to **add**. You can say **3 count on 2** or **3 add 2**.
How else can you say **3 count on 3**?
Now start on 8. Count back 2 places. What number are you on?
Counting back helps you to **subtract**.
You can say **8 count back 2** or **8 take away** 2.

Practise

1 Count on 2 each time.

a 4 add 2 is equal to ☐

b 5 add 2 is equal to ☐

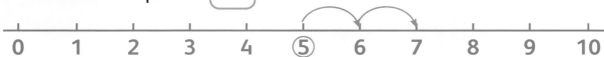

2 Count back 2 each time.

a 8 take away 2 leaves ☐

b 7 take away 2 leaves ☐

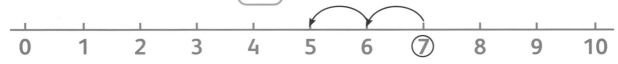

3 Use a counter on the number track to help you add or subtract.

| 1 | 2 | 3 | 4 | 5 | 6 | 7 | 8 | 9 | 10 |

a Add 2.

b Add 3.

c Add 4.

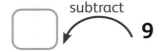

d Take away 3.

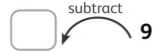

e Take away 4.

f Take away 5.

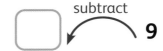

Adding numbers by counting on

Maths words

total   larger

Look at these 2 groups of fish.
How will you find the **total** number of fish?

 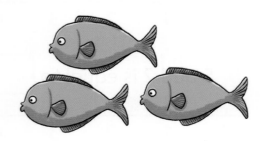

Here are 2 groups of shells.
How will you find the total number of shells?

Is there another way to find the total?

Learn

I have 5 cubes and 3 more cubes. What is the total?

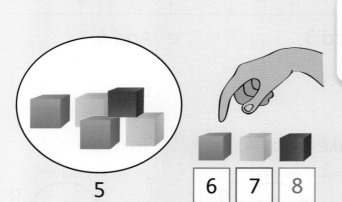

Start with the
larger number
and count on.

5 | 6 | 7 | 8 |

There are 8 cubes in total.
5 add 3 is equal to 8.

Practise

1   Use cubes. Count on to work out the total each time.

   a   4 add 3 is equal to ☐ cubes.

   b   4 add 4 is equal to ☐ cubes.

   c   4 add 5 is equal to ☐ cubes.

   d   3 add 2 is equal to ☐ cubes.

2   Count on to work out the total each time.

   a   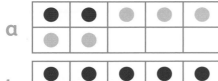   2 add 5 is equal to ☐

   b      8 add ☐ is equal to ☐

   c   ☐ add ☐ is equal to ☐

3   Maris scores 5 points in the first game.
   She scores 4 more points in the next game.
   How many points does she score in total?

Try this

What is 4 add 3?

Work with a partner and take turns.
Start with 5 cubes each. Pick a few more.
Find the total by counting on.
What number will you count on from?
Keep finding totals with different
numbers of cubes.

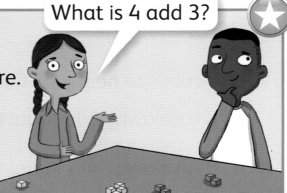

Subtracting numbers by counting back

Explore

The squirrels eat the same number of nuts from their piles.

How many nuts could each squirrel have left?

Learn

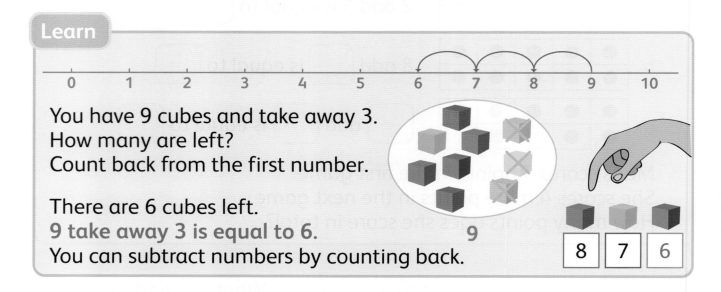

You have 9 cubes and take away 3.
How many are left?
Count back from the first number.

There are 6 cubes left.
9 take away 3 is equal to 6.
You can subtract numbers by counting back.

Practise

1 Use cubes to help you. Count back to find how many are left.

   a 8 take away 3 is equal to ☐ cubes.

   b 8 take away 4 is equal to ☐ cubes.

c 8 take away 5 is equal to ☐ cubes.

d 7 take away 2 is equal to ☐ cubes.

e 7 take away 3 is equal to ☐ cubes.

f 7 take away 4 is equal to ☐ cubes.

2 Count back on the ten frames.

a  7 take away 2 is equal to ☐

b 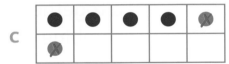 10 take away 4 is equal to ☐

c  6 take away ☐ is equal to ☐

3 David has 8 crayons.
He gives 3 crayons to his brother.
How many crayons does David have left?

Play *Squirrels and nuts* with a partner.
Start with 10 cubes each. 1 cube is 1 nut.
In turns, take 1, 2 or 3 nuts from your pile.
Count back to find how many nuts are left.
How many nuts are left after 3 turns?
Now start with 9 nuts in your pile, then 8, and so on.

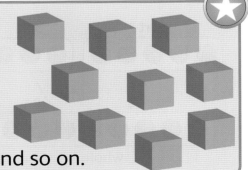

Finding totals

Explore

Make totals of 6 people in different ways.

Maths words
altogether
combine
part
whole

Learn

How many balls **altogether**?
Combine them and add.

part

part

whole

5 and 3 make a total of 8.

part

5

part

3

8

whole

Practise

1 Add the cubes in each row and write a **number sentence**.

a [ ] and [ ] make [ ]

b [ ] and [ ] make [ ]

c [ ] and [ ] make [ ]

d [ ] and [ ] make [ ]

2 Add the purple dots to find each total.

a [ ] • dots

b [ ] • dots

c [ ] • dots

d [ ] • dots

Maths word
number sentence

3 What is the total each time?

(3) (2) (5)

a (3) (3) ( )

b (3) (4) ( )

c (3) (5) ( )

d (4) (5) ( )

e (5) (5) ( )

Try this

How many counters to make each total? Find 3 examples.

 and make 8

 and make 10

23

Taking away

Maths word
left over

Explore

Each child wants a balloon, a cupcake, an apple and a drink.
Are there enough? How many of each will be **left over**?

Learn

There are 5 pies on the plate.
David takes away 3 pies.

Take away 3 pies from 5 pies;
2 are left on the plate.

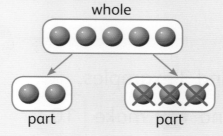

How many pies are left?

5 take away 3 leaves 2.

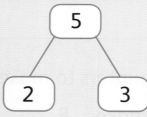

Practise

1 Complete each number sentence.

a 6 take away ☐ leaves ☐

b 5 take away ☐ leaves ☐

c 7 take away ☐ leaves ☐

d 8 take away ☐ leaves ☐

2 Use counters to help you answer these.

a 7 take away 1 ⟶ ☐   b 7 take away 3 ⟶ ☐

c 7 take away 5 ⟶ ☐   d 7 take away 7 ⟶ ☐

3 Write the missing number each time.

a 5 ○ 3   b 6 ○ 2   c 7 ○ 1   d 9 4 ○

Try this ★

How many pencils must you take away each time, so that only 1 pencil is left in each pot?

a 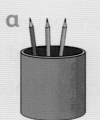   b    c 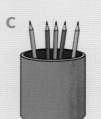   d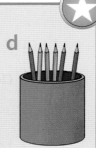

What patterns do you notice? Try to explain them.

Using the + and − symbols

BUS STATION

Bus 7 Bus 4 Bus 6 Bus2 Bus3

Use the bus numbers to make some addition sentences.
Now make some take away sentences.
Which number will you use first?

Learn

We can write **4 add 2 is equal to 6** as **4 + 2 = 6**.
We can write **6 take away 2 is equal to 4** as **6 − 2 = 4**.

The = symbol means **equals**.

Maths word
equals

Practise

1 Use + and = to write each addition sentence.

   a 4 add 3 is 7   b 5 add 3 is 8   c 6 add 3 is 9   d 5 add 1 is 6

2 Use − and = to write each subtraction sentence.

   a 7 take away 3 is 4   b 8 take away 3 is 5
   c 9 take away 3 is 6   d 6 take away 1 is 5

Practise (continued)

3 Complete the number sentences. Use the number line to help you.

```
  0    1    2    3    4    5    6    7    8    9   10
```

a 3 + 1 = ⬚

3 + 2 = ⬚

3 + 3 = ⬚

b 8 − 1 = ⬚

8 − 2 = ⬚

8 − 3 = ⬚

c 5 + 4 = ⬚

9 − 4 = ⬚

6 + 3 = ⬚

Try this

Work out each addition and subtraction.
Write your answers in boxes like those below.

5 + 2 = ⬚

7 − 4 = ⬚

8 − 3 = ⬚

3 + 1 = ⬚

9 − 2 = ⬚

2 + 3 = ⬚

Answer is less than 5	Answer is 5	Answer is more than 5

Let's talk

Look at your answers to the **Try this** activity.
Talk about other additions and subtractions to write in each box.

Comparing numbers

Explore

This street is busy. Are there more bicycles or cars?
How many more? What else can you **compare**?

Maths words
compare
longer
difference

Learn

The 7 blue cubes show the number of bicycles in the **Explore** picture.

The 5 orange cubes show the cars.

We can compare the number of cubes to find out how many more.

2 more

The line of blue cubes is **longer**.

Each line has 5 cubes. There are 2 more of the blue cubes.
We say **7 is 2 more than 5**, so the **difference** between 7 and 5 is 2.

Practise

1 Compare the rows of shapes.
   Say how many more each time.

   6 is 2 more than 4

   > We can also say:
   > The difference **between** 6 and 4 is 2.

   a ⬜ is ⬜ more than ⬜

   b ⬜ is ⬜ more than ⬜

   c ⬜ is ⬜ more than ⬜

   d ⬜ is ⬜ more than ⬜

2 Use the word **difference** to compare
   the numbers in question 1.

   Maths word
   between

Let's talk

Zara and David choose a number each.
David's number is 3 more than Zara's.
Which number did each child choose?
Find more than 1 solution.

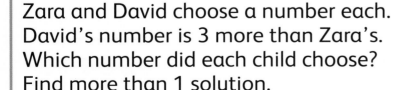

| 3 | 2 | 5 |
| 10 | 8 | 7 |

29

Quiz

1 Write true or false each time.
   a 6 is 1 more than 5.   **b** 9 is 1 less than 8.
   c 7 is 1 more than 5.   **d** 3 is 1 less than 4.
   e 10 is 1 more than 9.

2 Use the number tracks to help you.

   a 6 take away 4 equals ☐ | 1 | 2 | 3 | 4 | 5 | 6 | 7 | 8 | 9 | 10 |

   b 7 add 2 equals ☐ | 1 | 2 | 3 | 4 | 5 | 6 | 7 | 8 | 9 | 10 |

3 What is the total each time?

   a 5   1   **b** 6   4   **c** 2   7   **d** 4   3

4 Write addition sentences for question 3. Use + and =.

5 How many are left each time?

   a    **b**    **c**

6 Write subtraction sentences for question 5. Use − and =.

7 How many more each time?

   a    **b**    **c**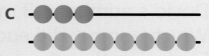

2D and 3D shape names

What shapes can you see?

How many **edges** on each?

How many **faces**?

What shapes are the faces?

Look at each **2D shape** below. 2D shapes are flat shapes.

A **circle** has 1 **side** that is **curved**.

A **triangle** has 3 **straight** sides and 3 **corners**.

A **square** has 4 straight sides of the same length and 4 corners.

A **rectangle** has 4 straight sides and 4 corners.

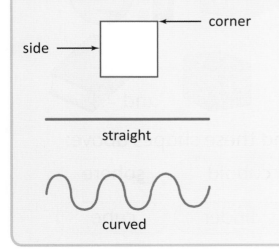

side ⟶

⟵ corner

straight

curved

Maths words	
edge	face
2D shape	circle
side	curved
triangle	straight
corner	square
rectangle	

Learn

Look at each **3D shape** below. 3D shapes are solid shapes.

A **sphere** has 1 curved surface.

corner ·····▶

A **cube** has 6 **faces**, 12 edges and 8 corners.

edge

face

A **cuboid** has 6 faces, 12 edges and 8 corners.

A **cylinder** has 2 faces, and 1 curved surface.

A **cone** has 1 face, 1 curved surface and 1 corner.

Maths words
3D shape
sphere
cube
face
cuboid
cylinder
cone

Practise

1

3

2

Find these shapes above:

(circle) (square)

(rectangle) (triangle)

Find these shapes above:

(cuboid) (sphere)

(cylinder) (cube)

Practise *(continued)*

3 What is each shape?

My shape has
3 corners and 3 sides.

My shape has 2 faces
and 1 curved surface.

Try this

Look around the classroom.
How many different shapes
do you see?

How will you record them?
Record means to keep track of.

Let's talk

Talk to your partner.

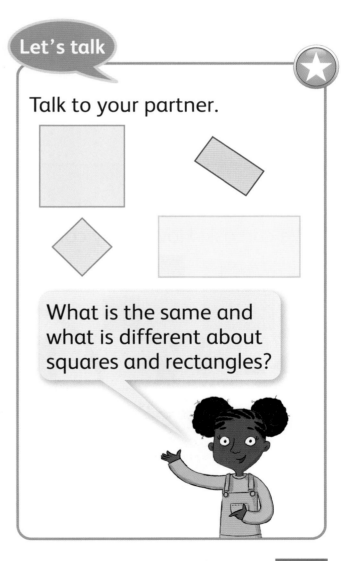

What is the same and
what is different about
squares and rectangles?

Sorting shapes

Explore

How have the clothes been sorted?
Can you **sort** them in a different way?

Maths word
sort

Learn

We can sort shapes in different ways.
We can ask: How many corners? How many sides?

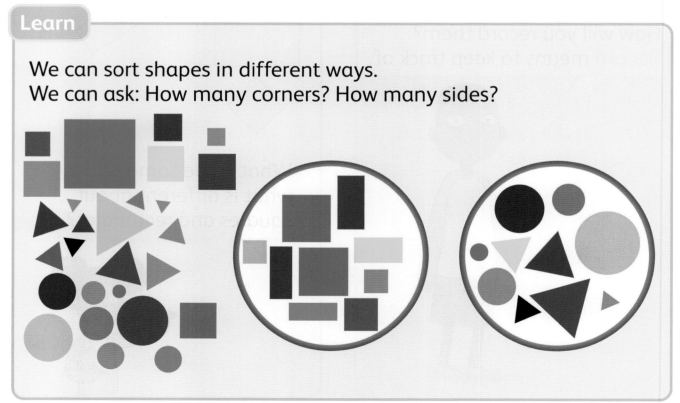

Practise

1 Sort the shapes into circles like these.

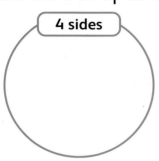 4 sides

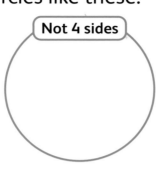 Not 4 sides

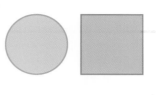

2 Sort the objects into a table like this.

Curved surfaces	Not curved surfaces

 cylinder

 pyramid

 sphere

cube

cuboid

Try this

Sort the objects in as many ways as you can.

Let's talk

Work with a partner to think of a heading for each circle.

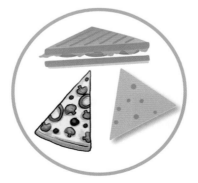

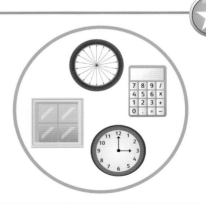

Patterns and pictures

Explore

What pictures and models can you see?

What shapes can you see?

Maths word
pattern

Learn

We can make **patterns** by repeating shapes.

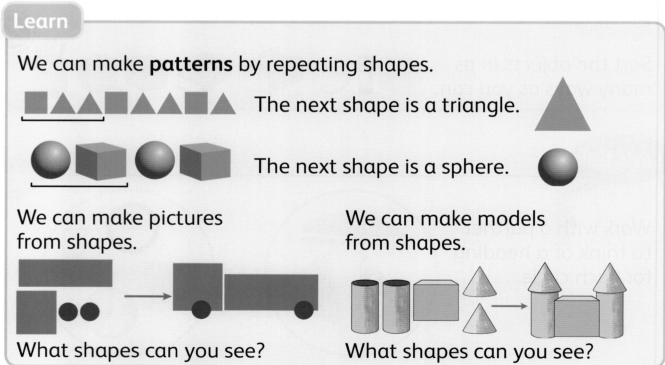

The next shape is a triangle.

The next shape is a sphere.

We can make pictures from shapes.

We can make models from shapes.

What shapes can you see?

What shapes can you see?

Practise

1 Draw the next 2D shape for each pattern.

a

b

2 What is the next 3D shape in each pattern?

a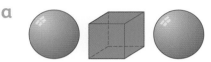

b

c

d

3 What shapes are in each picture?

a

b

c

Try this

Use shapes like these to create a picture.

Let's talk

Zara has made a necklace. What shapes did she use?

Quiz

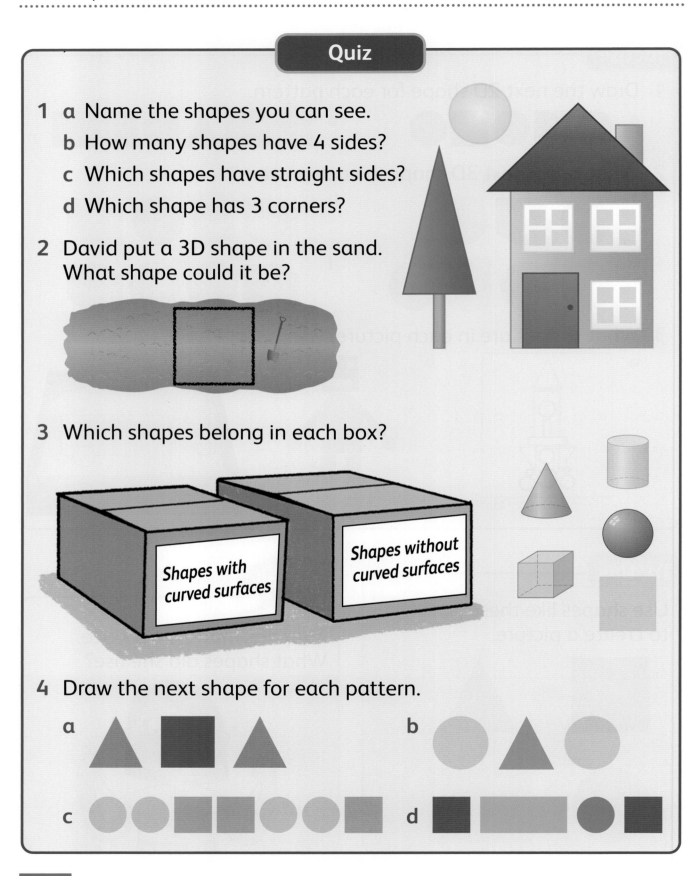

1 a Name the shapes you can see.
  b How many shapes have 4 sides?
  c Which shapes have straight sides?
  d Which shape has 3 corners?

2 David put a 3D shape in the sand.
  What shape could it be?

3 Which shapes belong in each box?

Shapes with curved surfaces

Shapes without curved surfaces

4 Draw the next shape for each pattern.
  a
  b
  c
  d

Statistical methods

Tables, lists and pictograms

Explore

Jack and Annay looked for different shapes.

What shapes did Annay find?

How many squares did he find?

Which shape did Annay find most of?

What other way could you show the information?

How could you **organise** the information?

Learn

We can write a **list**, for example:

Circle ● – clock, tomato, plate	clock
Square ■ – window, tissue	tomato
Rectangle ▬ – door, tray	plate
Triangle ▲ – pizza slice	window
	tissue
	door
	tray
	pizza slice

Maths words
organise
list

Learn (continued)

We can make a **table**.

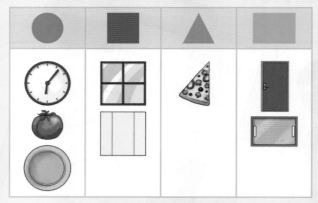

We can draw a **pictogram**.

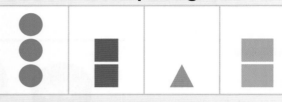

Which shape did Annay find the most of?

Which shape did Annay find the least of?

Practise

I drew the shapes I found!

1 Write a list of the shapes on Jack's clipboard.

2 Draw each shape Jack found.
  Use a table like this.

Circle	Square	Rectangle	Triangle

3 a Use the information or **data** that Jack collected in a pictogram like this.

A pictogram to show the shapes Jack found

Circle	● ● ●
Square	
Rectangle	
Triangle	

b How many circles did Jack find?

c Of which shape did he find the most?

d Of which shape did Jack find the least?

e How many shapes did he find altogether?

Maths words
table
pictogram
data

Venn diagrams and Carroll diagrams

Explore

David has sorted his toys.

What are the rules for each sorting hoop?

How else can you sort the toys?

Maths words
Venn diagram
Carroll diagram

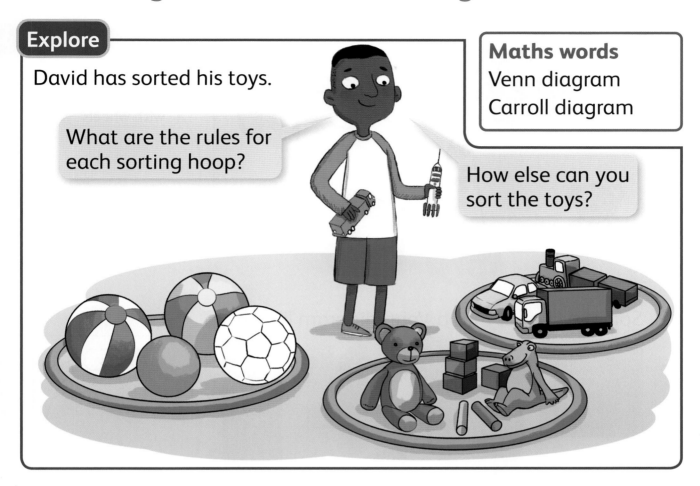

Learn

Venn diagrams

This is a **Venn diagram**.

3 sides

Where does this shape go?

Carroll diagrams

This is a **Carroll diagram**.

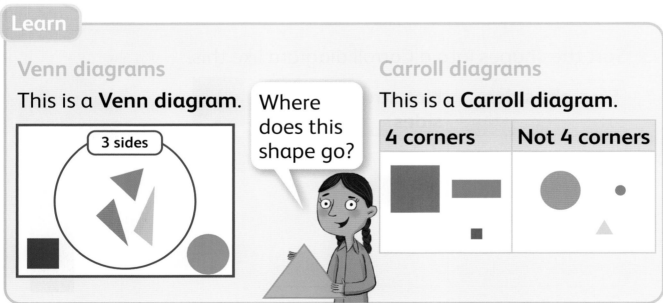

4 corners	Not 4 corners

Practise

1   Sort the objects into a Venn diagram like this.

Objects
with wheels

2   Sort the shapes into a Venn diagram like this.

4 sides

3   Sort the shapes into a Carroll diagram like this.

Straight sides	Not straight sides

Block graphs

Explore

Name each fruit you see.
Can we show this information
in other ways?

Learn

We can draw a table.

How many
bananas?

Fruit		Number of fruits
Mango		4
Apple		6
Banana		5
Watermelon		2
Coconut		1

Learn (continued)

We can draw a **block graph**.

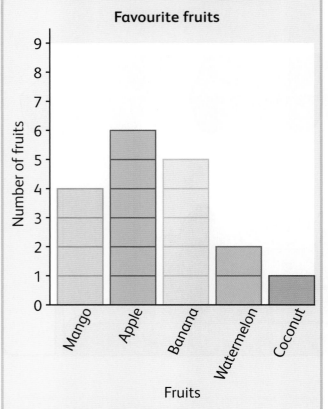

Favourite fruits

How many mangoes?
How many fruits altogether?
Are there more mangoes
than apples?

Let's talk

Share your favourite
sports with a partner.

Maths word
block graph

Practise

1 Draw a block graph to show
the **data** in the table,
about favourite sports.

Favourite sport	Number of people
Football	4
Cricket	3
Hockey	2
Swimming	6

2 Look at the block graph.

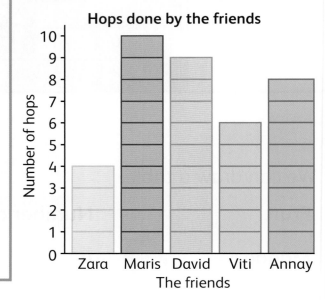

Hops done by the friends

a How many hops did Annay do?
b Who did 9 hops?
c Who did the most hops?
d Who did the least hops?
e How many hops did Viti and Zara
do altogether?
f Did Maris do less hops than David?

Quiz

1 Draw a pictogram to show the marbles.
   a How many red marbles are there?
   b What colour are there most of?
   c What colour are there least of?
   d How many marbles altogether?

2 Draw a Venn diagram like this one.
   a Draw two 2D shapes
      that go inside the hoop.
   b Draw two 2D shapes that
      go outside the hoop.

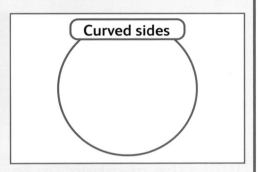

Curved sides

3 Look at the block graph.
   It shows favourite animals
   of some people.
   a How many people liked
      elephants best?
   b How many people liked
      tigers best?
   c How many people in total?
   d Which was the
      favourite animal?
   e Which was the least
      favourite animal?

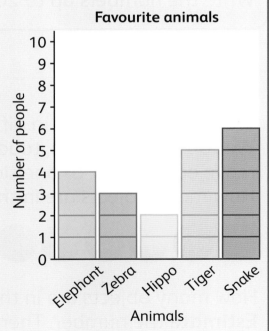

Favourite animals

Numbers to 20

How many?

Reading and writing numbers

What numbers did someone write in the sand? Say them.

Write the numbers up to 20 in sand or rice.

We can count objects in different ways.
We can point to them, touch them or
move each object to one side.
How many objects are in this row?

Maths word
estimate

How many objects are in the row below?
Estimate the number. Then count to check.

Practise

1 Count the rectangles. Make each number with cubes.

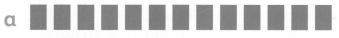

a

b

c

2 How many each time?

a

b

 3 Try to find these objects in your classroom.

a 13 pencils   b 15 books   c 18 shoes

Try this

Write the numbers from 1 to 20.
See how many numbers you can write without lifting your pencil off the page!

Let's talk

How many balls do you think there are?
Estimate first, then count to check.
How do you know when to stop counting?

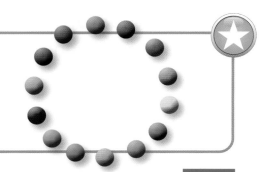

Counting on and counting back

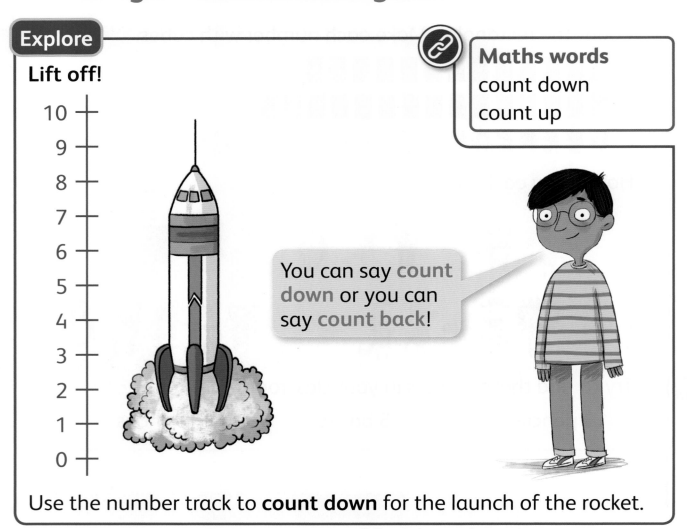

Explore

Lift off!

Maths words
count down
count up

You can say count down or you can say count back!

Use the number track to **count down** for the launch of the rocket.

Learn

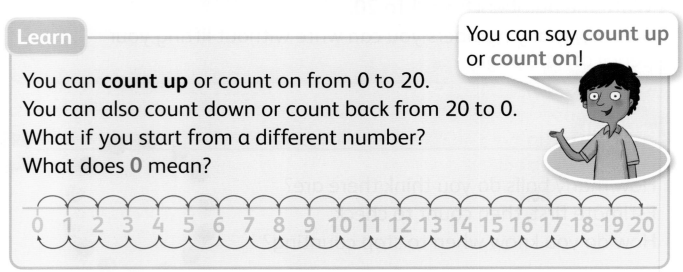

You can say count up or count on!

You can **count up** or count on from 0 to 20.
You can also count down or count back from 20 to 0.
What if you start from a different number?
What does 0 mean?

0 1 2 3 4 5 6 7 8 9 10 11 12 13 14 15 16 17 18 19 20

Practise

1 Complete the pattern by making the towers and writing the numbers.

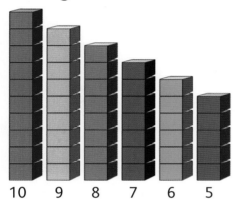

| 10 | 9 | 8 | 7 | 6 | 5 |

Use the number track below for question 3.

2 Look at the animal track. What number is each animal on? Count on. Then count back to check.

Start

1 2

7

12

15

Finish

18 20

3 a Start on number 9.
   Count up from 9 to 20.
   b Start on number 16.
   Count back from 16 to 1.
   c Choose your own starting number.
   Count up or back on the number track.

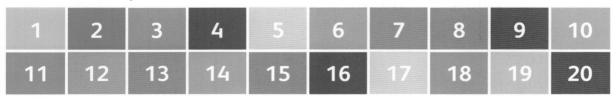

1	2	3	4	5	6	7	8	9	10
11	12	13	14	15	16	17	18	19	20

Try this

Choose any flag to start. Follow the flags and count up to 20.
What numbers are missing from the flags?

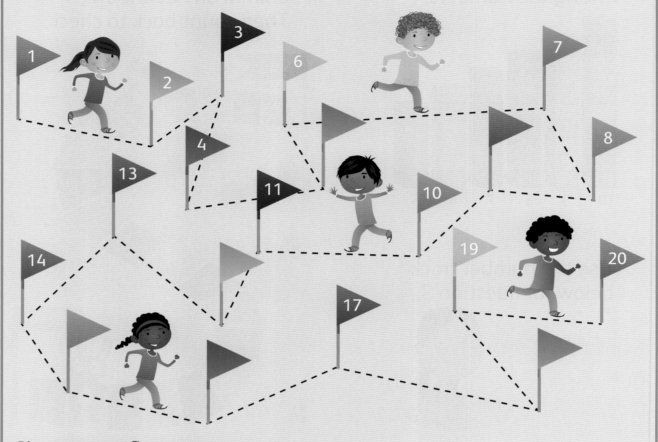

Choose any flag to start.
Follow the flags and count down to 1.

Let's talk

What does the number **0** mean?
Show **zero** fingers.
Show zero toys.
Show zero elephants!

Maths word
zero

Larger and smaller numbers

Explore

Who used more blocks?

Maths words
taller   larger
smaller

Mine is **taller**. I used more blocks.

Learn

Count the balls in each group. Which group has more?
13 comes after 12, so it is a **larger** number.
12 is the **smaller** number.

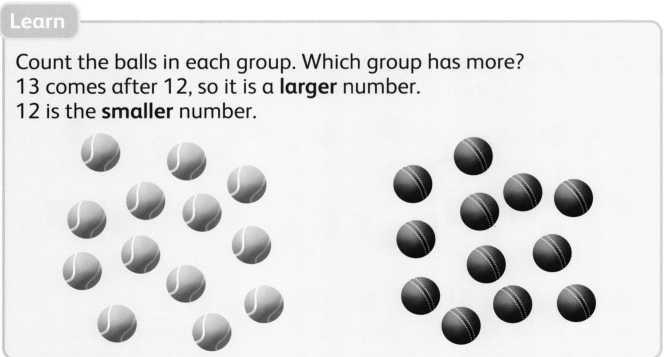

Practise

1 Count the dots.
Which number is larger? Which number is smaller?
Use the number track to help you.

| 1 | 2 | 3 | 4 | 5 | 6 | 7 | 8 | 9 | 10 | 11 | 12 | 13 | 14 | 15 | 16 | 17 | 18 | 19 | 20 |

a

b

c

d

2 Use cubes to make towers. Which tower is taller?

9 or 5 cubes?

The red tower
with **9** cubes
is taller.

a 8 or 10 cubes?
b 13 or 14 cubes?
c 20 or 19 cubes?
d 15 or 5 cubes?

3 Count each kind of shape. Which has more?
Estimate first, then count to check.

a

b

c

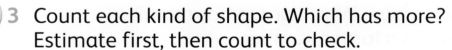

Try this

What could Annay's number be?
What is Jack's number?

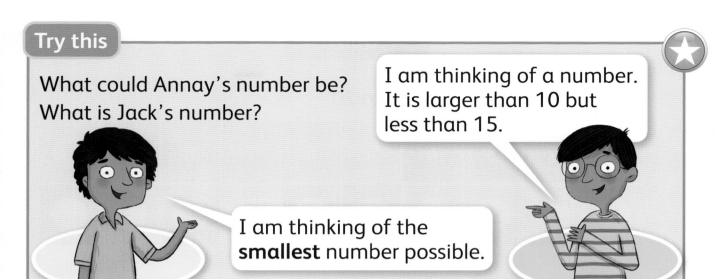

I am thinking of a number. It is larger than 10 but less than 15.

I am thinking of the **smallest** number possible.

Let's talk

How could you check if a number is larger?
Think of different ways to check.

Maths word
smallest

Quiz

1 Write the numbers from 10 to 20.

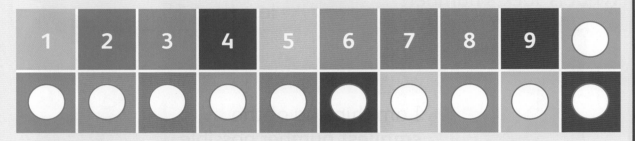

2 How many cubes?

3 Make a tower with 20 cubes.

4 Count on from 5 to 10.

5 Count down from 12 to 0.

6 Which set of counters has a larger number?

7 Which number is smaller?

14   9

Time and measurement

Time

Explore

What is Jack doing in the **morning**?

What is Jack doing in the **afternoon**?

What is your **daily routine**?

Maths words
morning
afternoon
daily routine

Learn

We use different words to describe the **time** of a **day**.

| morning | **noon** | afternoon | **evening** | **night** |

Some things take a **long** time. Some things are **quick**.
A baby becoming a child takes a long time.

> I took 8 years to grow to the height I am now.

We **measure** this **amount** of time in **years**.
David is 8 years old.

Washing your hands is quick.

> Washing your hands with soap and water is very important.

Maths words
time
day
noon
evening
night
long
quick
measure
amount
year
minute

We measure this amount of time in **minutes**.

Jack took 1 minute to wash his hands well.

Practise

1 What time of day is it in each picture?

day time   night time

a

b

c

d

e

2 Draw what you do at these times. Use a table like this.

Morning	Afternoon	Evening	Night

Practise *(continued)*

3   How long does each activity take?

[**years**]   or   [**minutes**]

a    b    c    d

4   Put these pictures in order, to show how the plant grows.

a 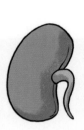   b    c    d

Try this

What can you do in 1 minute?

How many jumps?

How many times can you bounce a ball?

How many times can you clap your hands?

Let's talk

Use these words to tell a partner what you did yesterday.

[**morning**]   [**afternoon**]

[**evening**]   [**night**]

Length

Explore

How **tall** are the children?
Who is the **tallest**?
Who is the **shortest**?

Learn

We can compare lengths.
Zara is the shortest. Maris is the tallest.
Zara is **shorter** than Viti and Maris.
Viti is **taller** than Zara but shorter than Maris.
Maris is taller than Zara and Viti.

Maths words
tall
tallest
shortest
shorter
taller

Practise

1 Annay made a tower 7 cubes tall.
   a Make a shorter tower.
   b Make a taller tower.

2 a Which giraffe is the tallest?
   b Which giraffe is the shortest?

A   B   C

3 a Which snake is the longest?
   b Which snake is the shortest?

A

B

C

4 a Which snail has travelled the longest distance?
   b Which snail has travelled the shortest distance?

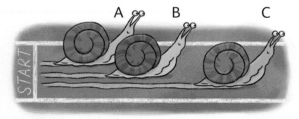

A   B   C

START

Maths words

longer   longest

Try this

Find as many objects as you can that are:
- **longer** than 5 cubes
- shorter than 5 cubes.

Let's talk

Which piece of string is the **longest**?
How do you know?

Quiz

1 Sort the activities below in a table like this.

A long time	A short time

a   b   c   d   e

2 Order these pictures. Start with **morning**.

a   b   c   d

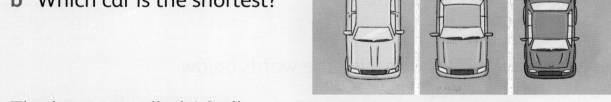

morning   evening   noon   afternoon

3 a Which car is the longest?
   b Which car is the shortest?

4 The learners rolled 4 balls.
   a Which ball travelled the longest distance?
   b Which ball travelled the shortest distance?

Units 1–6

1 Close your eyes. Your partner puts 1 to 10 counters in your hand. Count them with your eyes still closed. Swap.

2 Use the number line each time.

0 1 2 3 4 5 6 7 8 9 10

a Start on 6 and count on 3. What number are you on?

b Start on 8 and count back 4.
What number are you on now?

3 Complete.

a 7 + ☐ = 10   b 8 + ☐ = 10   c ☐ + 1 = 10

4 Name each 2D shape. Use the words below.

a   b   c   d

square   rectangle   circle   triangle

5 Name each 3D shape. Use the words below.

a   b   c 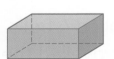   d

cuboid   sphere   cylinder   cube

6 Jack made a pictogram to show his friends' favourite animals.

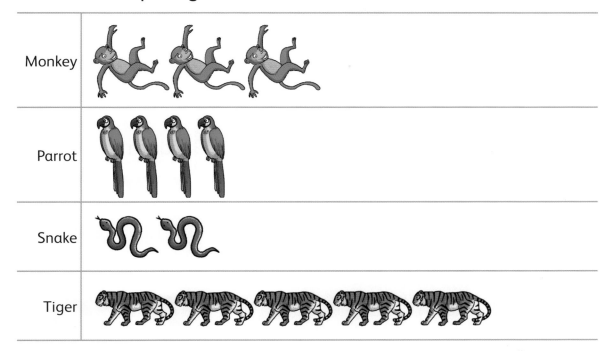

Monkey	
Parrot	
Snake	
Tiger	

a How many friends liked monkeys best?

b How many friends liked parrots best?

c Which animal did his friends like the most?

d Which animal did they like the least?

e How many friends did Jack ask altogether?

7 a Count up from 0 to 20. Miss a number on purpose.
Ask your partner to tell you which number you missed.

b Try the same game.
This time count down from 20 to 0.

8 a Which is the shortest tree?

b Which is the tallest tree?

Pictograms and block graphs

Explore

Maths word
data

What is your favourite flavour of ice cream?

How many chocolate ice creams do you see?

What is your least favourite flavour of ice cream?

Can you show the **data** in different ways?

Data is another word for **information**.

Learn

You can show the data as a **table**.

Flavour of ice cream	Number of ice creams
Chocolate	4
Strawberry	2
Coconut	1
Mango	3

You can show the data as a **pictogram**.

A pictogram to show favourite flavours of ice cream

Chocolate	
Strawberry	
Coconut	
Mango	

You can show the data as a block graph.

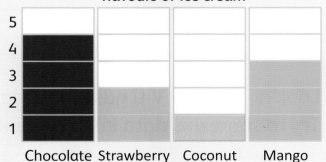

A block graph to show favourite flavours of ice cream

Chocolate Strawberry Coconut Mango

How many mango ice creams do you see?
Which ice cream was the favourite?
How many ice creams altogether?
Compare the pictogram and the block graph.
What is the same? What is different?

Practise

1 Annay and Jack looked for bugs.

Worms	
Spiders	
Ants	
Beetles	

a How many worms did they find?
b Which bug did they see most often?
c Which bug did they see least often?
d How many bugs did they find altogether?

2 Show the data as a block graph.

Try this

Look at the way Viti has sorted her toys.
Try to show this data in a different way.

Venn diagrams and Carroll diagrams

How could you sort these clothes?

Learn

Venn diagrams

You could sort the clothes by **colour** in a **Venn diagram**.

You could also sort the clothes by the **numbers** on them.

Carroll diagrams

You could draw a **Carroll diagram**.

More than 5	Not more than 5
6, 7, 9	1, 2, 3, 4

Where would the number 8 go?
Try to think of another way to sort the clothes.

Maths words
Venn diagram
Carroll diagram

Practise

1 Sort the objects into a Venn diagram like this.

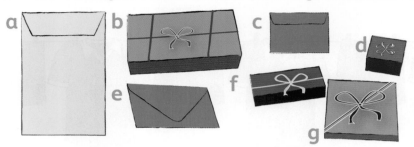
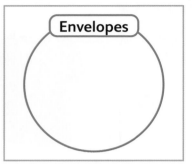

Envelopes

2 Sort the numbers into a Venn diagram like this.

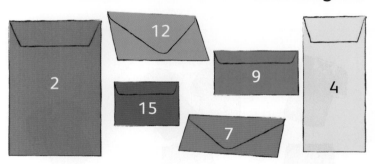
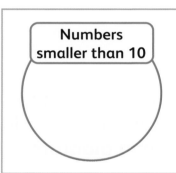

Numbers smaller than 10

3 a Sort the numbers into a Carroll diagram like this.

| 12 | 2 | 3 |

Less than 10	Not less than 10

| 6 | 17 | 14 |

   b Choose 2 more numbers to add to the Carroll diagram.

Try this

How could you sort these numbers?

| 4 | 5 | 12 | 16 | 18 | 3 |

Block graphs

Explore

Annay has done his skips.
Now he is writing the data for his friends.

Name	Number of skips
...ay	8
Zara	5
Viti	9
David	6
Maris	

__ , 5, 6, 7

Could you show the data in a different way?
Who did the least skips?
How many skips altogether?

You could draw a block graph to show the skipping data.
How many skips did Maris do?

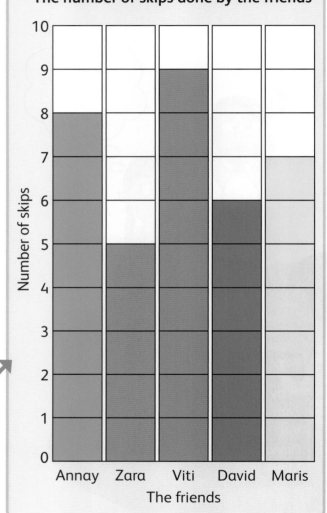

The number of skips done by the friends

Number of skips

Annay   Zara   Viti   David   Maris

The friends

What other questions could you ask about this block graph?

The friends did another round of skipping.

This time Annay did 9 skips, Zara did 8, Viti did 5, David did 8 and Maris did 7 skips.

1   Draw a block graph to show this skipping data.

2   Who did the most skips?

3   Who did the least skips?

4   Who did the same number of skips?

5   How many skips did Maris and Zara do altogether?

See how many skips you and 2 friends can do. Draw a block graph to show your data.

How will you collect the data?

Quiz

1 Draw a pictogram to show the flowers.

2 Draw a block graph to show the flowers.

3 a How many yellow flowers?

 b What colour of flower has the most?

 c What colour has the least?

 d How many flowers altogether?

 e How many more red flowers than purple flowers?

 f Write a question about the flowers. Ask your partner.

4 Use a Venn diagram like this.

 a Write 2 numbers that go inside the hoop.

 b Write 2 numbers that go outside the hoop.

Less than 10

5 Answer the questions about this Carroll diagram.

6, 9, 3, 7	17, 18, 12, 11

 a What headings could you use for this Carroll diagram?

 b Where would you put number 5?

 c Where would you put number 14?

8

Shapes, direction and movement

Naming and sorting shapes

Explore

What shapes can you see?
What is the same about the shapes?
What is different about them?

Learn

2D shapes

circle

| 1 curved side |

triangle

| 3 straight sides, 3 corners |

square ■

| 4 straight sides of equal length, 4 corners |

rectangle ▬

| 4 straight sides, 4 corners |

pentagon ⬠

| 5 straight sides, 5 corners |

hexagon ⬡

| 6 straight sides, 6 corners |

 curved  straight corner — side

We can sort shapes in different ways. See below:

Flat 2D shapes

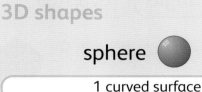

Maths words

pentagon   hexagon   pyramid

3D shapes

sphere

| 1 curved surface |

cube

| 6 faces, 12 edges, 8 corners |

cuboid

| 6 faces, 12 edges, 8 corners |

cylinder

| 2 faces, 1 curved surface |

cone

| 1 face, 1 curved surface, 1 corner |

pyramid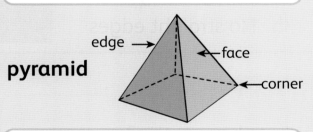

edge → face ← corner

| 5 faces, 5 corners |

Solid 3D shapes

Practise

1 Name each 2D shape.

circle | triangle | rectangle | square | pentagon | hexagon

a   b   c   d   e   f

2 Name each 3D shape.

sphere | pyramid | cube | cuboid | cone | cylinder

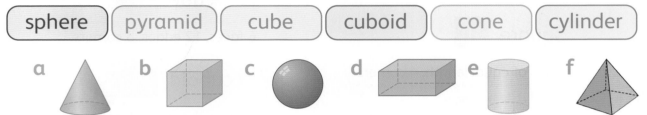

a   b   c   d   e   f

3 **a** Sort these shapes into a Venn diagram like this.

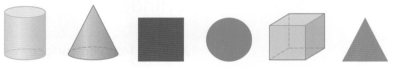

> 3D shapes

    b Name 2 shapes with a curved face.

4 Sort these shapes into the 2 groups. Use a Carroll diagram.

   a Straight edges

   b No straight edges

Try this

What shape is this?
Make other shapes.
How many pencils
do you need to
make a square?

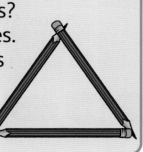

Let's talk

What is the
difference between
a 2D shape and
a 3D shape?

Patterns and pictures

What do you see in this picture?

What shapes can you see?

We can repeat shapes to make patterns.

The next shape is a circle.

The next shape is a triangle.

We can use shapes to make pictures.

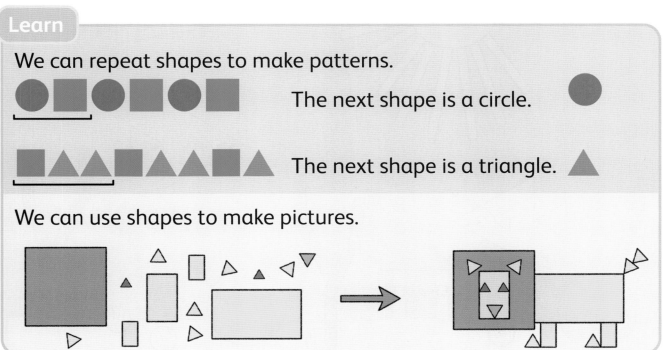

Practise

1 Draw repeating patterns using **2D** shapes.

   a Use circles and squares.

   b Use triangles and squares.

   c Use squares, triangles and circles.

> Try making different patterns using the same shapes.

2 Make repeating patterns using **3D** shapes.

   a Use cylinders and cubes.

   b Use cubes and cones.

   c Use cubes, pyramids and cuboids

3 What shapes can you see in this picture?

Position and direction

Explore

Where is the starfish?
Do you see a spotty fish?
What else do you see?

Maths words
up
down
left
right
below
turn
forwards
backwards

Learn

People, animals and objects can move in different ways.
Look at the arrows.

up	down
left	right
below	turn
forwards	
backwards	

Which way is each
arrow pointing?

Learn *(continued)*

We can use different words to describe **position**.

The starfish is **under** the bridge.

The blue fish is **above** the bridge.

The red fish is **in front** of the spotty fish.

The spotty fish is **behind** the red fish.

Practise

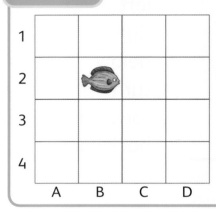

Start in square **B2** each time.
Follow the directions.
Which square does the fish swim to?
1 1 square forwards
2 1 square backwards
3 2 squares forwards
4 2 squares down

Maths words

under	above
in front	behind

Try this

Your partner must move like a robot. Give safe instructions.

Turn left, turn right, walk forwards 2 steps, walk backwards 1 step.

Keep your eyes open. Only do what your partner says if it feels safe. Take steps slowly and carefully.

Let's talk

When might you need to give instructions in everyday life?

Quiz

1 Name each shape below. Use the words to help you.

hexagon | triangle | pentagon | square | pyramid | cube | cone

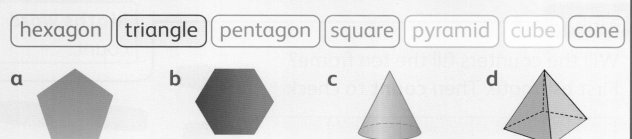

a b c d

2 Sort the shapes below into a table like this.

2D shapes	Not 2D shapes

3 Use shapes like this to draw a picture.

4 Are they moving to the left or to the right?

a b c

9

Numbers to 20

Ten and ones – making numbers

Explore

Will the counters fill the ten frame?
First estimate. Then **count** to check.

Maths word
count

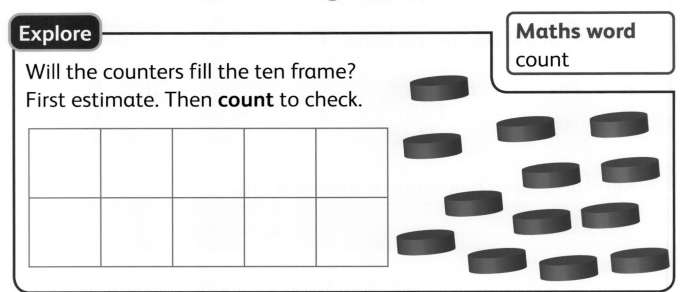

Learn

How many counters?
How can you tell?

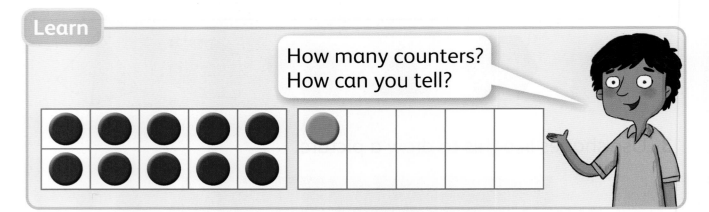

Practise

1 a Put 13 counters in ten frames like these.

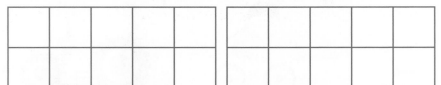

   b Now use 19 counters.

Practise *(continued)*

2   How many each time? Estimate. Then count to check.

a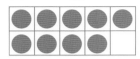

There are ☐ circles.

b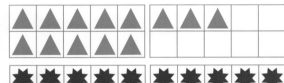

There are ☐ triangles.

c

There are ☐ stars.

Try this

Viti made 15 from 10 and 5.
Make 17.
Make other numbers from
10 and ones.

Let's talk

Who is right? Annay or Jack?

I made 14.

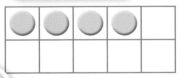

No. 10 must be first.

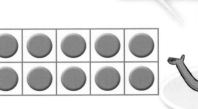

Ten and ones – breaking up numbers

This group of children **break up** into 2 **parts**.

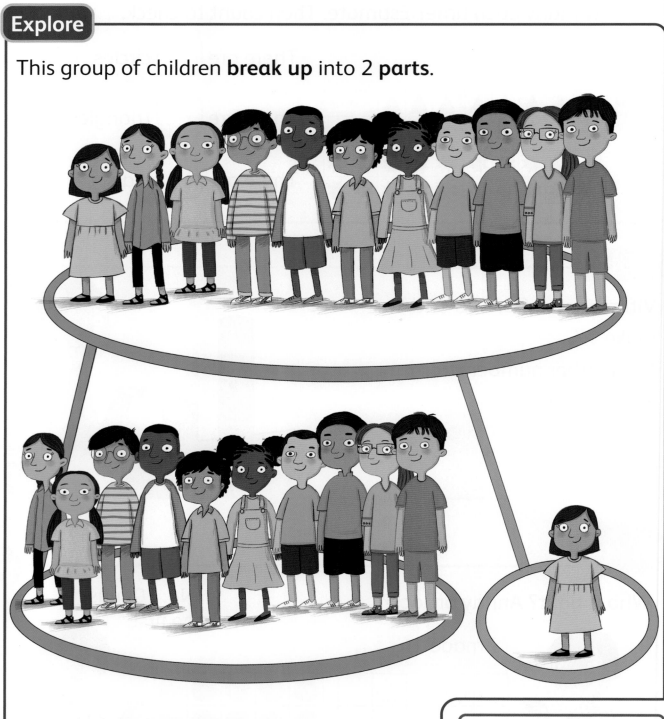

What is the whole number?
What are the numbers in each part?

Maths words

break up   part

Learn

We can break up 16 into 1 ten and 6 ones.

$16 = 10 + 6$

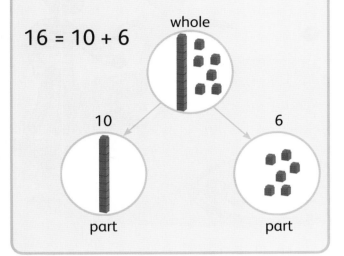

Practise

1 Use cubes to break up each number into 10 and ones.

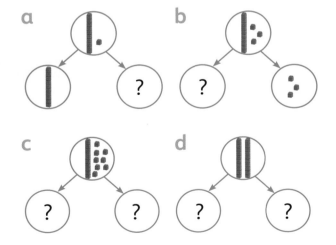

2 Draw a picture of this:
Break up 15 into 10 and ones.

3 Break up these numbers into 10 and ones.

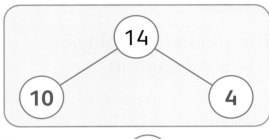

a
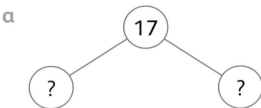

b
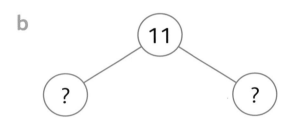

c
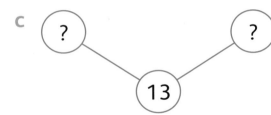

d
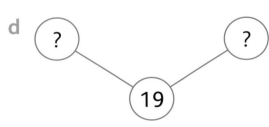

e
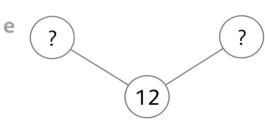

Comparing and ordering numbers

Explore

The children are playing
Throw a beanbag.

Who is winning?
How can you compare the 2 scores?

Learn

Put these numbers in order,
starting with the smallest.

| 16 | 6 | 11 |

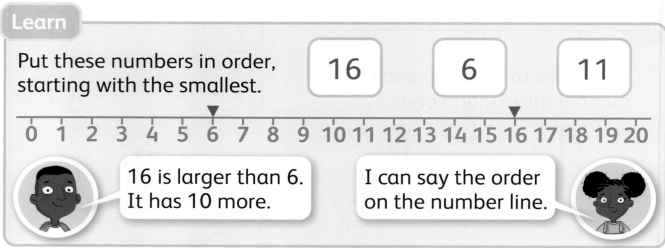

16 is larger than 6.
It has 10 more.

I can say the order
on the number line.

Practise

1 Use cubes to make the numbers
in each group. Put them in order.

a 15, 7, 14, 10   b 17, 13, 12, 16
c 18, 9, 11, 20   d 19, 8, 10, 6

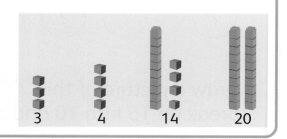

Practise *(continued)*

2 Write the numbers in order.

a

b

c

d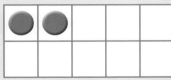

3 a Your teacher will clap and tap. Listen.
Did you hear more claps or taps?

b Write your name. Count each letter. How many?
Compare with friends. Who has the longest name?

Try this

Put these numbers in order from largest to smallest.
You need to work out 2 of the numbers first.

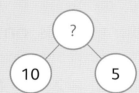

Let's talk

Talk about different ways to compare
the numbers.

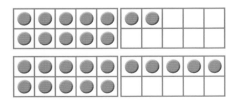

Counting on 10. Counting back 10.

Play the 10s counting game!
Start on a yellow square.
Count on 10.

You can start on
any yellow square.

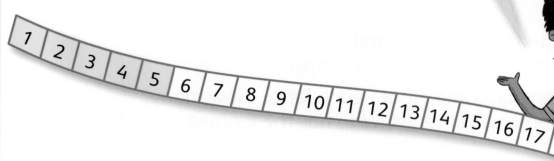

Say your start number. Then say your end number.

Learn

Count in 10s, on and then back.

1	2	3	4	5	6	7	8	9	10
11	12	13	14	15	16	17	18	19	20

Count on 10 from 3. Count back 10 from 13.
What do you notice? Use ten frames to show this.

Practise

1 Count on 10 from the number in the circle each time.

a

1	②	3	4	5	6	7	8	9	10
11	12	13	14	15	16	17	18	19	20

b

1	2	3	4	5	⑥	7	8	9	10
11	12	13	14	15	16	17	18	19	20

c

1	2	3	4	5	6	7	⑧	9	10
11	12	13	14	15	16	17	18	19	20

2 Use the number line below.
  a Count back 10 from 11.
  b Count back 10 from 15.
  c Count back 10 from 19.

0 1 2 3 4 5 6 7 8 9 10 11 12 13 14 15 16 17 18 19 20

3

I am thinking of a number.

My number is 10 more than yours.

What could Annay and Jack's numbers be?

Try this

Count on 10 from 0. Count on another 10.
What numbers did you land on? Try to explain this.

Quiz

1 What is the number each time?

a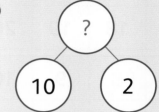

b
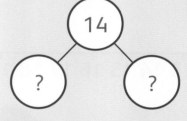

c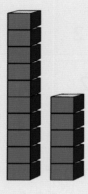

2 Break up each number into 1 ten and ones.

a   b 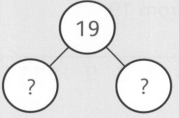   c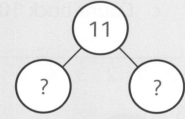

3 Say these numbers in order from smallest to largest.

18   5   15   8

4 Count on 10 from 7.

5 Count back 10 from 16.

# 10 Time and measurement

Time

# 10 Time and measurement

Time

Maths words
week   after   before

Explore

Maris and David went to a shop. They saw a sign on the door.

Monday – Closed
Tuesday 9:00 to 6:00
Wednesday 9:00 to 6:00
Thursday 9:00 to 8:00
Friday 9:00 to 8:00
Saturday 9:00 to 8:00
Sunday – Closed

Which 2 days of the **week** is the shop closed?
How many days of the week is the shop open?
What day comes **after** Friday? What day comes **before** Friday?

Learn

There are 7 days in a week.

There are 12 **months** in a year.

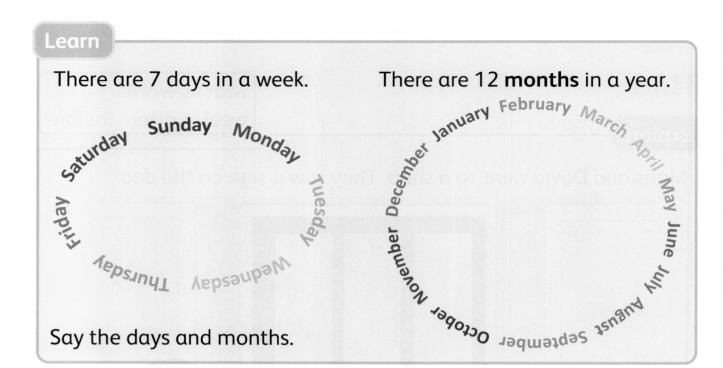

Say the days and months.

Practise

Write the days of the week in order.

| Saturday | Friday | Thursday |

| Tuesday | Monday | Wednesday | Sunday |

Maths word
month

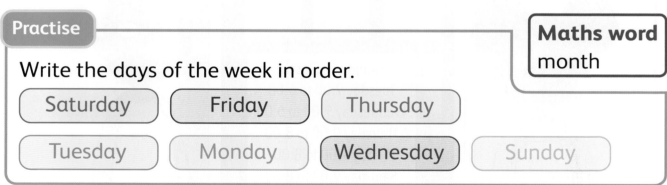

Try this

David meets his friend on Wednesday. As he is leaving, Annay says:

See you tomorrow, David.

What day of the week will Annay see David?

Let's talk

Make up a song or rhyme to help you remember the days of the week.

Length

What do you see in the picture?

Maths words

longest   shorter   shortest
longer   measure   length
long

Which snake is the **longest**? Which toucan has the **shorter** beak?

Learn

We can compare lengths.
The green snake is longest.
The blue snake is **shortest**.
The yellow snake is **longer** than the blue snake.
The yellow snake is shorter than the green snake.
We can use cubes to **measure** the **length** of an object.
The ribbon is 5 cubes **long**.

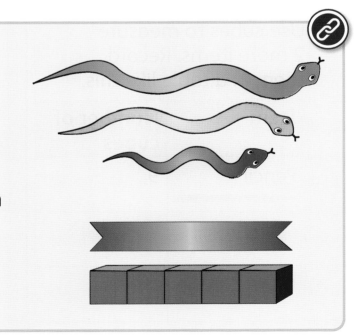

Practise

1 Zara made a snake.
It is 6 cubes long.

   a Make a shorter snake.

   b Make a longer snake.

2

How many cubes long
is the pencil?

3 How many
cubes long
is the book?

4 Use cubes to measure
2 more items. Record
them in a table like this.

Item	Number of cubes
Pencil	9

Try this

Put your hand on a sheet
of paper. Stretch your fingers
as wide as you can.
Draw around your hand.
Measure your
hand span
with cubes.

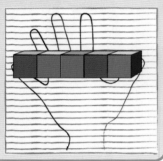

Let's talk

Is Maris correct?
How do you know?

The blue pencil is 8 counters
long. The red pencil is 3 cars
long. I think the blue pencil
is longest. 8 is larger than 3.

Mass

Explore

These books are so **heavy**.

My bag is much **heavier**.

Which do you think is **heaviest**?
How could we compare the objects?

Maths words

heavy	heavier
heaviest	mass

Learn

We can compare **mass**.

Is the pineapple heavier than the orange?

We find the mass of items by using a balance.

The pineapple has a mass of 9 cubes.

The orange has a mass of 3 cubes.

Practise

1 What is the mass of each parcel?

a    b    c

2 What is the mass of each fruit?

a    b    c

3

David finds the mass of some of his things.

a Which object is the **lightest**?

b Which object is the heaviest?

c Which object is **lighter** than the car?

d Which object is heavier than the car?

Maths words
lightest
lighter

Try this

Maris' parcel is heavier than David's parcel.
David's parcel is heavier than Annay's parcel.
Which parcel is Annay's?
What is the mass of David's parcel?

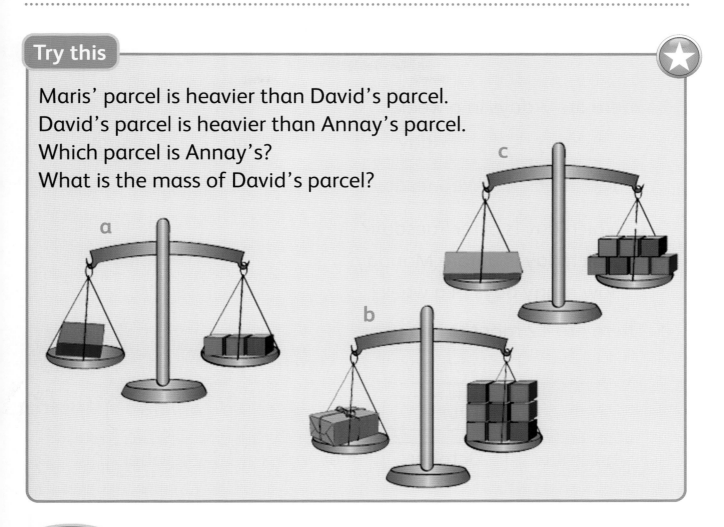

Let's talk

Who has the heaviest parcel?
How could Zara and Viti
find out?
Explain your answer.

My parcel is
larger, so it is
the heaviest.

Quiz

1 There are 7 days in a week.

| Saturday | Friday | Thursday | Tuesday |

| Monday | Wednesday | Sunday |

a Which day is after Tuesday?

b Which day is before Monday?

c How many days in a week?

2 There are 12 months in a year.

| January | February | March | April |

| May | June | July | August |

| September | October | November | December |

a Which month is after January?

b Which month is before November?

c How many months in a year?

3 Measure the pencil
in cubes or counters.

start   finish

4 What is the mass
of the orange?

Addition facts to 5

Explore

Look at my addition facts!

I think we need to check them.

$$1 + 1 = 2 \qquad 1 + 2 = 3$$

$$2 + 1 = 4 \qquad 2 + 2 = 5$$

$$3 + 1 = 4 \qquad 3 + 2 = 5$$

$$4 + 1 = 5$$

Help Annay and David to check the addition facts.
Are there any patterns that could help them?

Learn

1 + 1 = 2 is an addition fact.

We can use this fact to find other facts. When we make the first number 1 larger, the **total** is 1 larger.

When we make the **second** number 1 larger, the total is 1 larger again.

■ + ■ = ■ ■
1 + 1 = 2

■ ■ + ■ = ■ ■ ■
2 + 1 = 3

■ ■ + ■ ■ = ■ ■ ■ ■
2 + 2 = 4

Practise

1 Use cubes to make new number facts for each question. Write the new fact each time.

   a Make the first number 1 larger: ■ ■ + ■ ■ = ■ ■ ■ ■

   b Make the second number 2 larger: ■ ■ + ■ = ■ ■ ■

2 Find the missing numbers.

   a 2 + 1 = ☐

   b 2 + 2 = ☐

   c 2 + 3 = ☐

   d 3 + ☐ = 5

   e 2 + ☐ = 4

   f ☐ + 2 = 3

Let's talk

☐ + ☐ = 4   ☐ + ☐ = 5

Maths words
total   second
more

Jack thinks there are many ways to make number facts for 4.
Annay thinks there are **more** ways to make number facts for 5.
What do you think? Prove your thinking in your own way.

Pairs that total 10

Explore

It's school outing day!

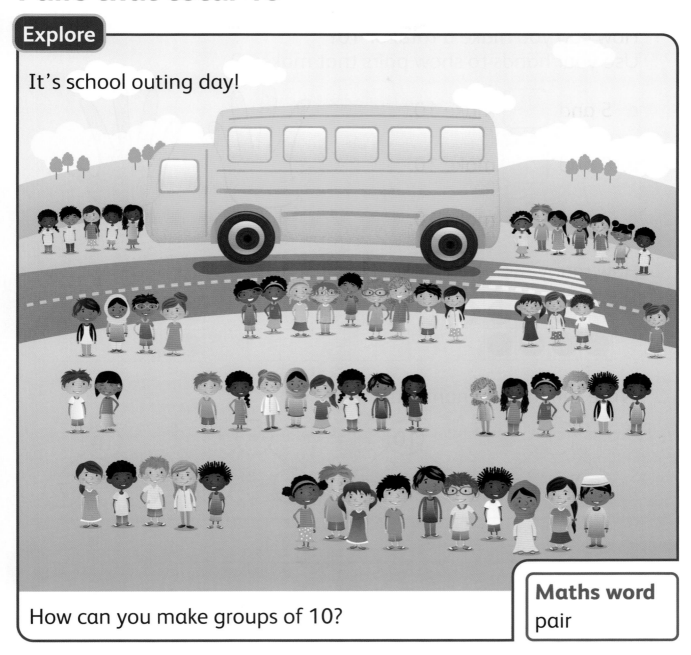

How can you make groups of 10?

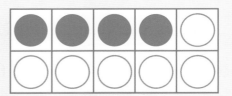

Maths word
pair

Learn

There are 4 blue counters.
How many **more** counters make 10?
The **pair** of numbers, 4 and 6, make 10.

Practise

1 How can you make a total of 10?
   Use your hands to show pairs that make 10.

   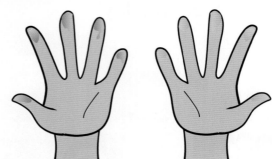

   a 5 and ⬜ make 10.

   b 8 and ⬜ make 10.

   c 3 and ⬜ make 10.

2 How many more will make a total of 10?

   a 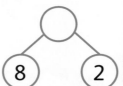 b   c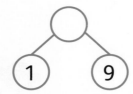

3 Write the missing numbers.

   a ◯ / 8 2   b 10 / ◯ 3   c ◯ / 1 9   d 10 / ◯ 7

Try this

Sort the addition pairs below into a table like this.

Total of pair is less than 10	Pair totals 10	Total of pair is greater than 10

4 and 6   5 and 4   5 and 6   7 and 1   3 and 7   10 and 0   3 and 8

Equivalent addition facts

Explore

Help Viti and David.

I need 5 balls in total for my game.

I need 4 balls in total for my game.

How can they get the balls they need?

Can they each take only 1 bag of balls?

Can they each take 2 different bags?

Maths word
equals

Learn

3 add 1 is equal to 4.
We can write this as:
$3 + 1 = 4$

2 add 2 also **equals** 4.
We can write this as:
$2 + 2 = 4$

Both addition facts above have the same total.

We can show this as:
$3 + 1 = 2 + 2$

The addition facts on each side of the equals sign have the same total.

1 Make the addition facts on each side equal.
How many balls do you need to put into the empty bags?

a  b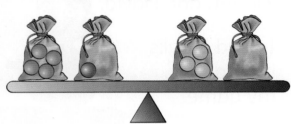

2 Use pairs that total 10 to make each side equal.

a 5 + ⬚ = 6 + 4   b ⬚ + 9 = 5 + 5

c 7 + 3 = 8 + ⬚   d 4 + 6 = ⬚ + 7

3 Are these number sentences true or false?

a 1 + 2 = 2 + 1   b 2 + 2 = 1 + 4   c 2 + 3 = 4 + 2

d 9 + 1 = 8 + 3   e 5 + 5 = 6 + 5   f 7 + 3 = 6 + 4

Try this

Maths word
amount

Maris has 7 cents in one pocket.
She has 3 cents in her other pocket.

> I have a different **amount** of money in each of my two pockets. I have the same amount of money as Maris in total.

How much money can Zara have in each pocket?

Show your solution as 7 + 3 = ⬚ + ⬚

Counting on for addition

Explore

Playing a game together is fun.

It's my turn first.
I can move from my
number 10 to the
same space as Zara.

It's my turn next.
I can move to 15.

On her next turn, Maris takes the lead.
Zara wins because her game piece lands on 20.
What numbers did they each spin in the game?

Learn

We can show counting on for addition on a number line.
This number line shows 10 + 3.

10 add 3 equals 13 because 10 count on 3 is 13.

We can also show 13 + 2 on the number line.

13 add 2 equals 15 because 13 count on 2 is 15.

Practise

1 Write the matching number sentences.

a

+ 2

0 1 2 3 4 5 6 7 8 9 10 11 12 13 14 15 16 17 18 19 20

b

+ 5

0 1 2 3 4 5 6 7 8 9 10 11 12 13 14 15 16 17 18 19 20

c

+ 5

0 1 2 3 4 5 6 7 8 9 10 11 12 13 14 15 16 17 18 19 20

2 Use the number line to count on 2 for each of these.

0 1 2 3 4 5 6 7 8 9 10 11 12 13 14 15 16 17 18 19 20

a 14 add 2 equals ☐

b 15 add 2 equals ☐

c 16 add 2 equals ☐

d 17 add 2 equals ☐

Let's talk

Look back at your answers for **Practise** question 2.
Use what you know to answer these questions.
- Is 13 add 2 more or less than 16?
- Is 15 add 3 more or less than 17?
- Is 16 add 4 more or less than 18?

Counting back for subtraction

Explore

It's Fun Day for the class!

Maths words
subtract
take away

Take turns. Choose a ball numbered 2, 3 or 5 from the bucket.

Each time, start on 20. Count back by the number on your ball.

The first player to get to a red flag wins.

Zara and Annay have 4 turns each. Only Zara gets back to a red flag.

Which ball numbers did Zara and Annay choose?

Learn

We can count back to **subtract** on a number line.
This number line shows 20 − 5.

Count back **5**

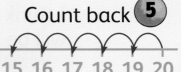

0 1 2 3 4 5 6 7 8 9 10 11 12 13 14 15 16 17 18 19 20

20 **take away** 5 equals 15 because 20 count back 5 is 15.

Do you think that 20 − 3 will be more or less than 15?
How do you know?

Practise

1 Use the number line to count back 2 each time.

0 1 2 3 4 5 6 7 8 9 10 11 12 13 14 15 16 17 18 19 20

  a 18 take away 2 equals ☐   b 17 take away 2 equals ☐

  c 16 take away 2 equals ☐   d 15 take away 2 equals ☐

2 Match each take away number sentence to its number line.

  a 13 take away 3 leaves 10.   b 13 take away 2 leaves 11.   c 14 take away 3 leaves 11.

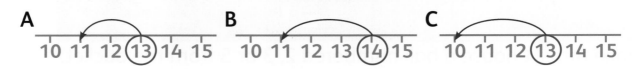

A 10 11 12 (13) 14 15   B 10 11 12 13 (14) 15   C 10 11 12 (13) 14 15

3 Are these number sentences true or false?
  a 13 take away 3 equals 10, so 13 − 4 is less than 10.
  b 16 take away 2 equals 14, so 16 − 3 is less than 14.
  c 20 take away 5 equals 15, so 19 − 5 is less than 15.

Try this

I think that 14 take away 3 will be less than 10.

Annay has made a mistake.
Explain the mistake to him.
Can you use another fact that you know to help?

Combining sets of objects to add to 20

Explore

Look at the orange and lemon trees.

How many baskets of fruit are on the table?

How many baskets of fruit are on the wall?

Are there more baskets of lemons or oranges in the picture?

How can you check?

Learn

Here are 6 bowls of oranges.

Here are 5 bowls of lemons.

We can use pairs that total 10 to help us add.
We know that 6 + 4 is 10, so 6 + 5 must be 1 more than 10.

6 + 4 = 10   1 more than 10 is 11.

We can see that 6 bowls is 1 more than 5 bowls.
So, we can add 6 and 5 in a different way.

5 + 5 = 10   1 more than 10 is 11.

Practise

1 How many bowls in total? Write number sentences.

a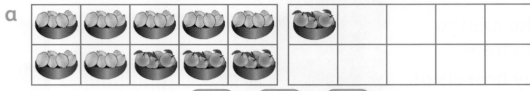

( ) + ( ) = ( )

b

( ) + ( ) = ( )

c

( ) + ( ) = ( )

2 Add 2 groups of fruit each time.
Write number sentences to match.

6   5   7   6

Let's talk

Look at your answer to Practise question 2.

Do any different additions give the same total?

Try to explain why.

Think about other additions that will have the same answer.

Taking away a small number of objects

Explore

There are lots of objects to choose from.

The children take turns to collect their objects.
How many objects are left each time?

Learn

The cubes in the ten frames show the 20 objects on the table in **Explore**.

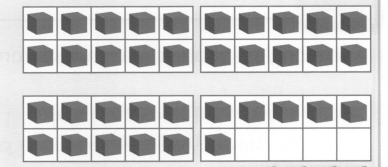

Jack takes 4 spinning tops.
So 16 objects are left.
We can write: 20 − 4 = 16

Zara takes 3 crayons.
So 13 objects are left.
We can write: 16 − 3 = 13

Viti takes 5 pebbles.
Will more than 10 objects be left on the table?
How do you know?

Practise

1   Take away 3 blocks each time.
    Say the subtraction number sentences.

a    b    c    d

2   Use cubes to find the answers to each set of subtractions.

   a   20 − 2 = ☐   20 − 3 = ☐   20 − 4 = ☐

   b   16 − 4 = ☐   16 − 3 = ☐   16 − 2 = ☐

Finding the difference

Here are some tower blocks in a model city.

How many more blocks are in the second tower than in the third tower?

Compare the number of blocks in the other pairs of towers.

How many more cubes are in the taller tower?

We need 2 more cubes to make the towers the same **height**.

The **difference** between 5 and 3 is 2.
5 − 3 = 2

We can check this by counting on a number line.
3 count on 2 is 5.

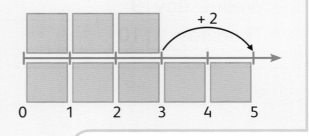

Maths words
compare   height
difference

Practise

1 Use cubes to copy each pair of towers.
What is the difference each time?

a   b   c   d

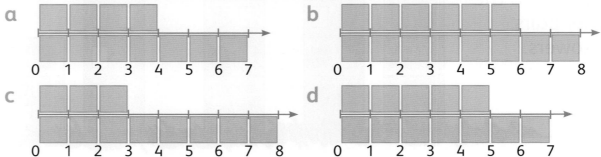

2 Write subtraction sentences to show the difference.

a

```
0  1  2  3  4  5  6  7
```

b

```
0  1  2  3  4  5  6  7  8
```

c

```
0  1  2  3  4  5  6  7  8
```

d

```
0  1  2  3  4  5  6  7
```

3 Work out the difference in morning and afternoon visitors for each day. Use a number line to help you.

Day	Morning visitors	Afternoon visitors
Monday	6	3
Tuesday	8	5
Wednesday	9	4
Thursday	10	2

Try this

The difference between 2 numbers is 3.
What could the 2 numbers be?
Find some pairs of numbers that make this true.
Think of a way to prove that your numbers are possible.

Quiz

1 Which pairs make 10?

2 Viti has 10 cubes in total.
6 cubes are in one cup.
How many cubes are under the other cup?

3 Use the number line to help you complete
the number sentences.

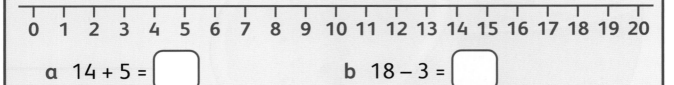

a 14 + 5 = ☐   b 18 − 3 = ☐

4 Viti has 12 red buttons and 5 blue buttons.
How many buttons in total?

5 Annay has 14 shells.
He drops 5 shells.
How many shells are left?

6 David has 9 stickers.
Maris has 5 stickers.
The difference between 9 stickers and 5 stickers is ☐ stickers.

Fractions are equal parts

Fruit is so good for you! Does fruit taste sweet or sour?

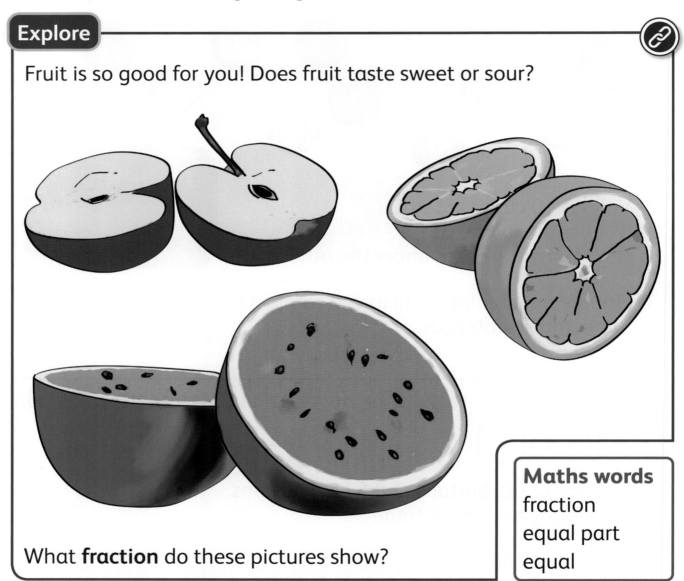

Maths words
fraction
equal part
equal

What **fraction** do these pictures show?

Learn

A fraction is an **equal part** of a whole.
Which circle shows equal parts?
Which circle does not show equal parts?

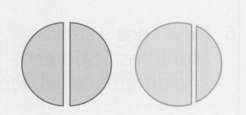

Practise

1 Point to the pictures that show equal parts.

a

b  c 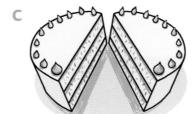 d

2 Draw a picture that shows 2 equal groups.

I drew equal parts to show fractions.

3 a Make a tower of 10 cubes.
   Break it up into 2 equal towers.
 b Make another tower of 10 cubes.
   Break it up into 2 unequal towers.
 c How many other ways can you do this?

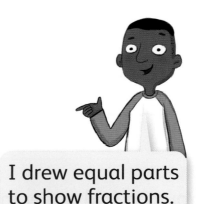

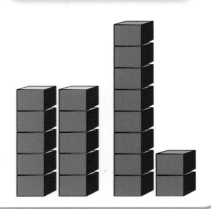

Try this

Take 12 cubes. Make 3 towers.
Try to make 3 **equal** towers.
Now try with 13 cubes. What do you notice?

Finding half of a shape

Explore

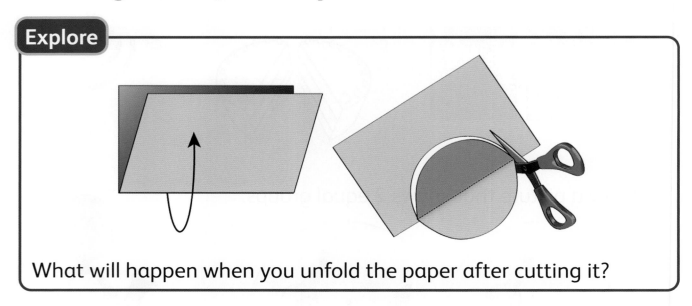

What will happen when you unfold the paper after cutting it?

Learn

Maths word
half

You can fold paper in **half**.
Which piece of paper shows half?

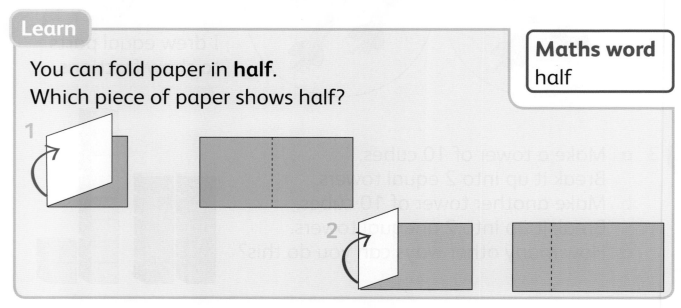

Practise

1 Fold a strip of paper in half.
Colour in 1 half.

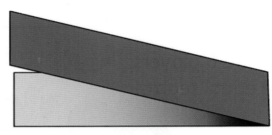

2 Take a square.
Fold it in half.
Try different ways.

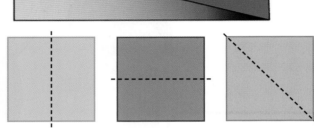

3 Which shapes show **half**?

a  b  c 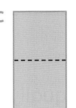 d  e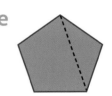

Try this

Make a butterfly.

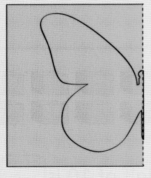 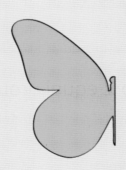 

Fold a sheet of paper or card in half.
Trace or copy the half butterfly.
Ask for help to cut out the butterfly carefully.
Open it out and flap the wings!

Equal sharing

Explore

There are 6 players.
Make equal teams.

Maths word
share

Learn

How can you **share** the cakes equally?

What is half of 8?

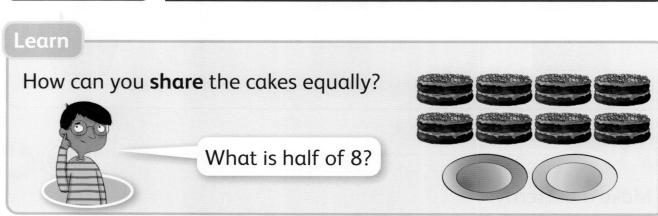

Practise

1 Take 8 cubes. Share them into 2 equal groups.
Make 2 equal towers.

2 Find half.

a Half of 4 is ☐

b Half of 6 is ☐

c Half of 10 is ☐

Practise (continued)

3   The hand shows half of the counters.
     How many counters in total?

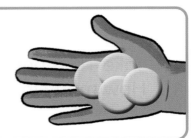

Try this ⭐

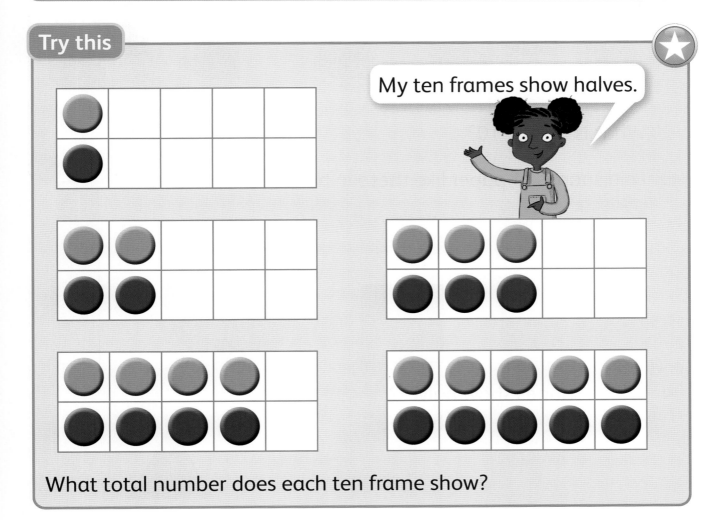

My ten frames show halves.

What total number does each ten frame show?

Let's talk ⭐

Can you share these sweets equally?
Why or why not?

Quiz

1 Take 10 cubes. Share them into 2 equal groups.
How many in each group?

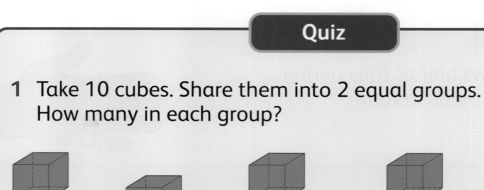

2 What is half of 10?

3 Fold shapes of paper like these in half.

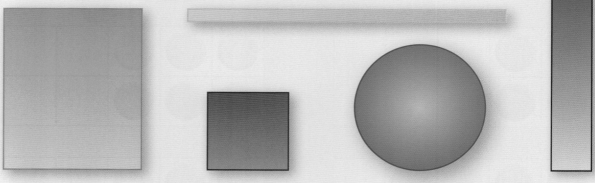

4 Which shapes show halves?

a 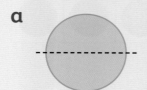 b 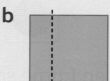 c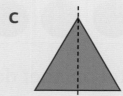

5 Which groups can you share to make equal parts?

a 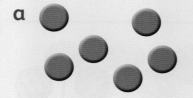 b  c

Units 7–12

1 Complete a Venn diagram like this.

 3   5   6

 8   10 15

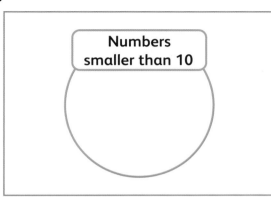

Numbers smaller than 10

2 Draw the next 2 shapes in this pattern.

3 Use 13 counters. Fill a ten frame.
How many counters will be left over?
Explain how you know.
Try the same with 17 counters.
What if you had 20 counters?

4 Write the days of the week in order.

Friday   Saturday

Thursday   Monday

Sunday   Wednesday   Tuesday

5 What is the mass of the parcel?

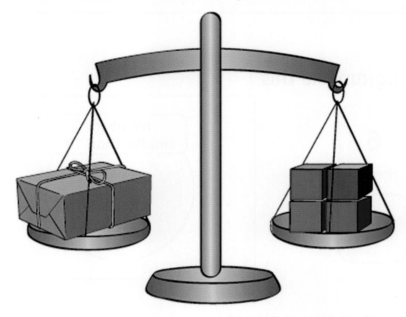

6 Find the missing numbers to make these number sentences true.

a 2 + ☐ = 4 + 1

b 3 + 3 = 4 + ☐

c 4 + 3 = 3 + ☐

d 5 + 3 = ☐ + 4

7 Find the difference each time.

a Between 6 and 2

b Between 9 and 5

c Between 7 and 4

8 Draw pictures and answer the question.

a Draw a shape broken up into equal parts.

b Draw a shape broken up into unequal parts.

c Which picture shows a fraction?

13 Numbers to 20

Understanding numbers to 20

Explore

Do Viti and Zara have enough?

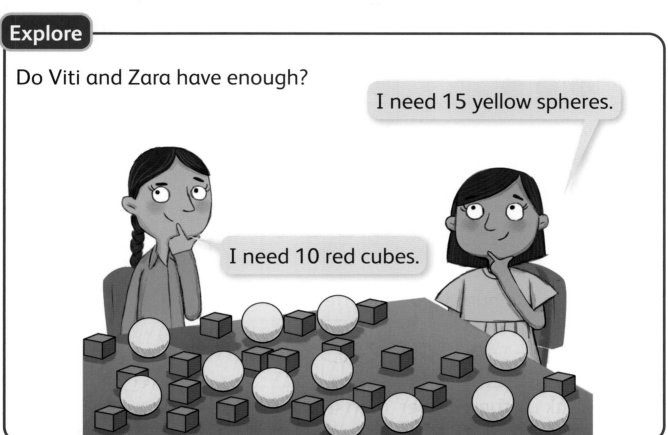

I need 15 yellow spheres.

I need 10 red cubes.

Learn

Each set shows how to make 14.

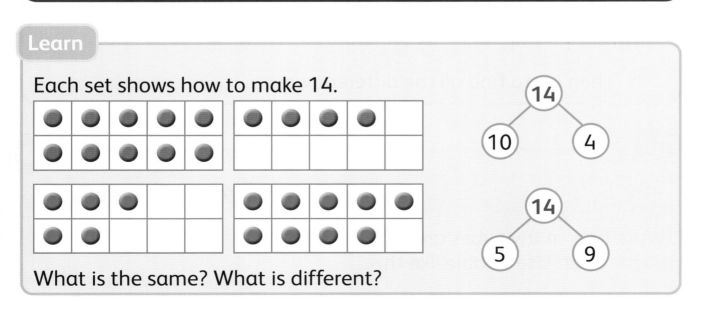

What is the same? What is different?

Practise

1   Write the missing part or parts.

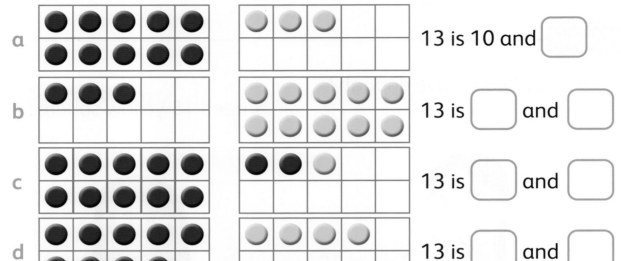

a   13 is 10 and [ ]

b   13 is [ ] and [ ]

c   13 is [ ] and [ ]

d   13 is [ ] and [ ]

 2   Break up the number **15** in different ways.

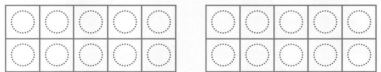

 3   a   Break up the number **11** in 4 more ways.

11 — 10   1
11
11
11
11

b   Then try to find all the different ways.

 **Try this**

Take 12 cubes.
Make 3 towers.
Write how many cubes are in
each tower. Use a table like this.

Tower A	Tower B	Tower C
8	3	1

First, second, third …

Explore

It's Race Day!

Who won?
Who came second?

Each learner has a number.
What do the numbers mean?

Maths words

first	second	third
fourth	fifth	

Learn

Look at this pattern.
Point to each shape and say the positions: 1st, 2nd, 3rd …

1st	2nd	3rd	4th	5th	6th	7th	8th	9th	10th
first	**second**	**third**	**fourth**	**fifth**	sixth	seventh	eighth	ninth	tenth

What shape is 1st? What shape is 4th?
What positions are the blue triangles?

Practise

1 Make a pattern of counters.
Say the order as you place them.

1st	2nd	3rd	4th	5th	6th	7th	8th	9th	10th

2

a What shape is 1st?

b What shape is 2nd?

c In what positions
   are the squares?

d What shapes would
   be 8th and 9th?

Try this

Say these patterns. Which comes 10th for each?

a

b

c

Let's talk

Teach a friend how to write your name.
What letter is 1st? What letter is 2nd? What letter is 3rd?

Even numbers and odd numbers

Explore

What is the same? What is different?

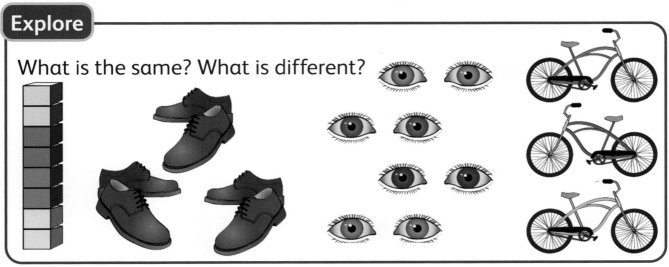

Learn

Maths words
even
odd

Start at 2 and count up in twos.

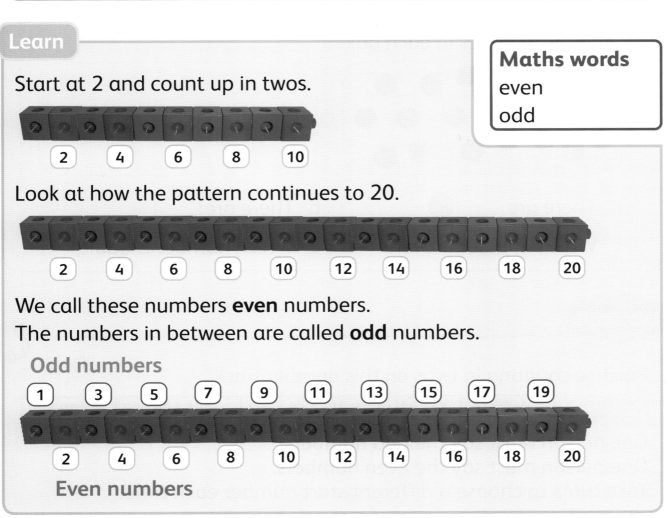

Look at how the pattern continues to 20.

We call these numbers **even** numbers.
The numbers in between are called **odd** numbers.

Odd numbers

1   3   5   7   9   11   13   15   17   19

2   4   6   8   10   12   14   16   18   20

Even numbers

Practise

1 Use cubes to make this tower. Count in twos.

2 Count in twos. How many are there?

a

b

3 Count the shapes in each box.

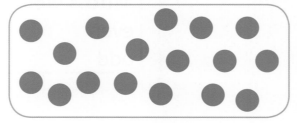
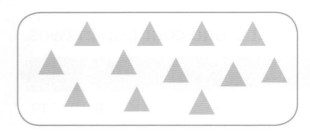

a There are ⬜ ●.

b There are ⬜ ▲.

c Which number is even?

d Which number is odd?

Let's talk

Practise counting in twos on this number track.

| ✗ | ② | ✗ | ④ | ✗ | ⑥ | ✗ | ⑧ | ✗ | ⑩ | 11 | 12 | 13 | 14 | 15 | 16 | 17 | 18 | 19 | 20 |

One person must say the odd numbers.
One person must say the even numbers.
Take turns to choose a different start number each time.

Money

Maths words

coin   cent

value

Explore

Does Annay have enough **coins**?

First estimate. Then count to check.

I need 20 coins.

Learn

What is the same about these coins?

What is different?

Which would you rather have? Why?

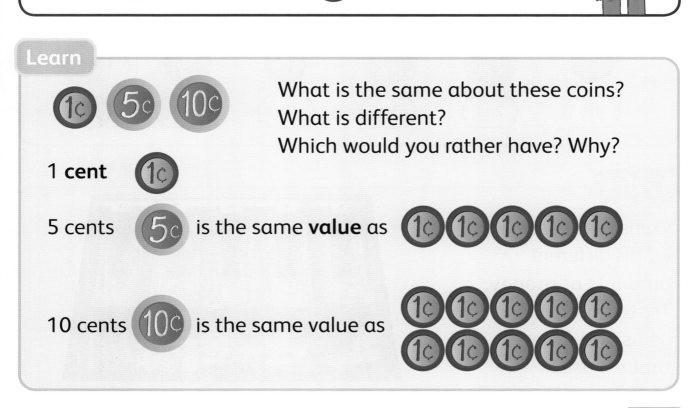

1 **cent**

5 cents   is the same **value** as

10 cents   is the same value as

Practise

1 How many coins?

a There are ☐ 1 cent coins.

b There are ☐ 5 cent coins.

c There are ☐ 10 cent coins.

d How many coins altogether?

> **Maths words**
> note
> money

2 Look at the pattern. What coin will be in 9th place?

1¢	5¢	10¢	1¢	5¢	10¢
1st	2nd	3rd	4th	5th	6th

Try this

Coins and **notes** are worth different amounts of **money**. Try to name all the coins and notes that we use.

Quiz

1 Show 2 different ways to break up **20** into parts.

2 These animals are standing in order in a row.
Answer the questions.

   a Point to the 4th animal.

   b Which animal is 1st?

   c In which position is the fish?

   d In which position is the horse?

3 Make a tower using an even number of cubes.

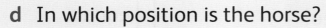

4 Make a stack of counters using any number you like.
Did you use an even number or an odd number of counters?

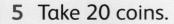

5 Take 20 coins.
   a Estimate how much money you have.
   b Sort the coins into piles of the same kind.
   c Which coin do you have most of?
   d Which coin do you have least of?

Addition facts to 10

Explore

It is a good day for drying clothes!

What addition facts can you see?

What is the **total** of each fact?

Maths words

total

altogether

equals

less than

more

Learn

The **Explore** picture shows 4 pink T-shirts and 2 yellow T-shirts.

There are 6 T-shirts **altogether**.

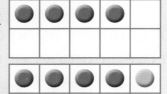

We can use the addition (+) sign and the **equals** (=) sign to write 4 + 2 = 6.

This tells us that 4 and 2 is equal to 6.

What is 2 + 4? What is 2 + 5? Which total is **less than** the other?

Try to explain why 2 + 5 is 1 **more** than 2 + 4.

Practise

1 Find each total.
   Write the matching addition sentences using + and =.

a 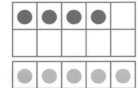   b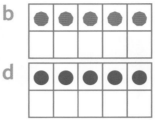

c 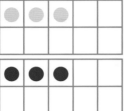   d

2 Work out the missing numbers.

a 3 + 1 = ☐
   1 + ☐ = 4
   3 + 2 = ☐
   ☐ + 2 = 5

b 4 + 2 = ☐
   4 + 3 = ☐
   ☐ + 3 = 8
   ☐ + 4 = 8

c 3 + 6 = 9
   ☐ = 3 + 6
   3 + 5 = 8
   ☐ = 5 + 3

3 a David puts 4 seeds in the first pot
     and 2 seeds in the second pot.
     How many seeds does he
     plant altogether?

  b Viti scores 5 goals in the first game
    and 3 goals in the second game.
    How many goals does she score in total?

Try this

Use cubes to **convince** your partner that these sentences are true.
5 + 4 is 1 more than 5 + 3.   6 + 3 is 1 less than 6 + 4.
4 + 4 is equal to 5 + 3.   6 + 4 is equal to 5 + 5.

Pairs that total 10

Explore

Making bracelets is fun!

The children are learning a new skill.

What patterns do you notice?

Learn

There are 10 counters in each ten frame.

1 and 9 make 10.
1 + 9 = 10

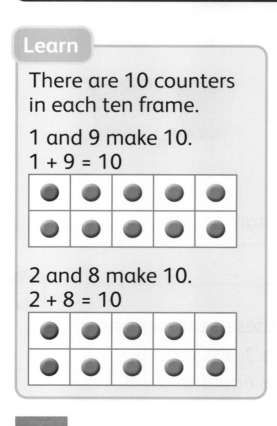

2 and 8 make 10.
2 + 8 = 10

Practise

1 Write an addition fact for each flower.

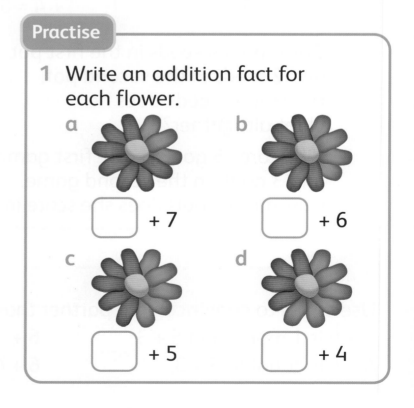

a

☐ + 7

b

☐ + 6

c

☐ + 5

d

☐ + 4

Practise *(continued)*

2 Work out the missing numbers.

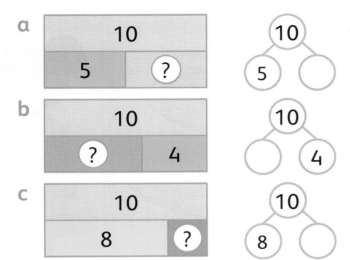

a

| 10 |
| 5 | ? |

10 — 5 — ○

b

| 10 |
| ? | 4 |

10 — ○ — 4

c

| 10 |
| 8 | ? |

10 — 8 — ○

3 Solve these number stories.

a There are 10 birds in my garden.
8 are in the tree.
The rest are on the wall.
How many birds on the wall?

b There are 10 ducks on the pond.
3 fly away.
How many ducks are left?

c There are 10 cars in the car park.
4 drive away.
How many cars are left?

Try this ⭐

Viti and Zara have
10 cents between them.
How many cents can
they each have?
How many different
answers can you find?

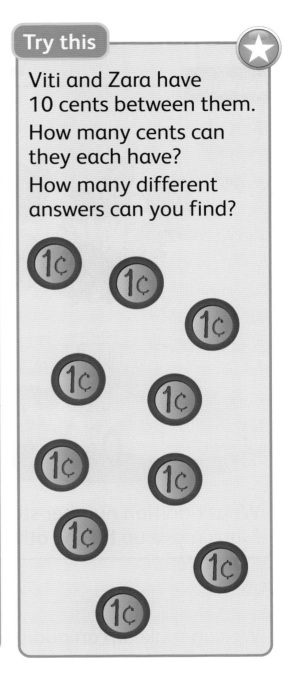

Let's talk

Share your answers to **Practise** question 3 with a partner.
Explain how you used pairs that total 10 to help you
to answer each number story.

⭐

Addition number stories

Explore

There are 5 apples on one plate and 3 on the other plate. There are 8 apples altogether.

What addition number stories can Maris make up for the other pictures?

Learn

We can make up an addition number story to match a number sentence.

For the number sentence, 4 + 2 = 6, we can say: *There are 4 eggs in a basket. Gran collects 2 more.*
Now there are 6 eggs in the basket.

What other number stories can you make up for the number sentence 4 + 2 = 6?

Practise

1 Write the letter of each number story and its matching
number sentence.

Number story   **Number sentence**

a

> There are 8 balloons.
> 7 balloons are red and
> 1 balloon is blue.

$5 + 3 = 8$

b

> There are 4 bats and the
> same number of balls.
> There are 8 toys in total.

$2 + 6 = 8$

c

> There are 2 crabs and
> 6 fish in a pool.
> There are 8 animals in total.

$7 + 1 = 8$

d

> Maris has 5 points.
> She scores 3 more.
> She has 8 points in total.

$4 + 4 = 8$

2 Make up your own number story for each addition sentence.

a $5 + 5 = 10$

b $7 + 2 = 9$

c $1 + 4 = 5$

Try this

Zara and Viti are making up number stories for an addition.

Do both number stories match 6 + 3 = 9?
How do you know?

There are 6 shells in my bucket.
I drop 3 shells.
There are 9 shells left in my bucket.

There are 6 boats on the river.
3 more boats come along.
There are now 9 boats on the river.

Adding on a number line

Explore

What will happen in the game?

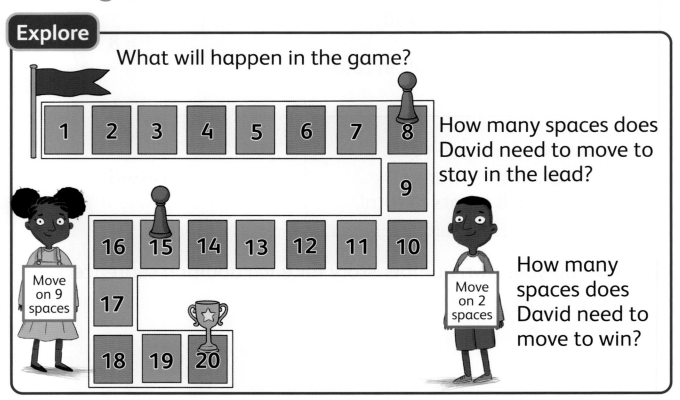

How many spaces does David need to move to stay in the lead?

How many spaces does David need to move to win?

Move on 9 spaces

Move on 2 spaces

Learn

Maths word
estimate

We can use a number line to help us add.
For 12 + 5, we can start on 12 and count on 5.

0 1 2 3 4 5 6 7 8 9 10 11 12 13 14 15 16 17 18 19 20

12 + 5 = **17**

What is 12 + 6?
Will it be more or less than 17? How do you know?
Making an **estimate** helps us to check answers.

Practise

1 Count on the number line to add the numbers.
Will your answers be more or less each time?

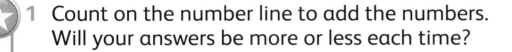

0 1 2 3 4 5 6 7 8 9 10 11 12 13 14 15 16 17 18 19 20

a 10 + 4
11 + 4
12 + 4
13 + 4

b 11 + 8
11 + 7
11 + 6
11 + 5

c 7 + 2
7 + 4
7 + 6
7 + 8

Will 11 + 4 be more
or less than 10 + 4?
Make estimates.

2 Zara uses the starting numbers. She spins a spinner and counts
on by that number. What number does she land on each time?

	a	b	c	d
Starting number	9	12	14	18
Spinner number	4	5	3	2
Lands on				

Practise *(continued)*

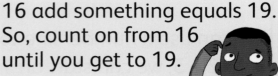

16 add something equals 19. So, count on from 16 until you get to 19.

3 Find the missing numbers.
Use the number line to help you.

16 + | 3 | = 19

0 1 2 3 4 5 6 7 8 9 10 11 12 13 14 15 16 17 18 19 20

a 11 + [ ] = 16   b 9 + [ ] = 13

c 14 + [ ] = 19   d 16 + [ ] = 20

Let's talk

I know that 5 + 5 = 10, so 6 + 5 will be more than 10.

I know that 8 and 9 are both very close to 10. So, 8 + 9 must be close to 20.

I know that 10 + 10 = 20, so 8 + 7 will be less than 20.

Work with a partner.
Use the children's thinking to estimate these additions.

4 + 5 = [ ]   7 + 6 = [ ]   12 + 11 = [ ]   8 + 8 = [ ]   9 + 7 = [ ]

Counting on from the larger number

Explore

There are lots of toys and books on the shelves.
Count on to find different totals.

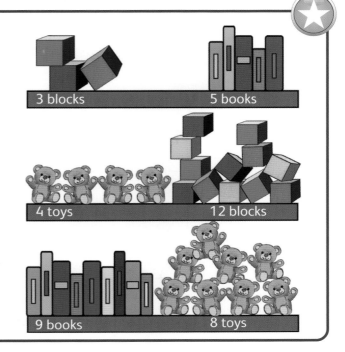

3 blocks   5 books

4 toys   12 blocks

9 books   8 toys

What number will you start on each time?

Learn

To add numbers together, we can count forwards.

I have 3 apples.

I have 16 apples. How many apples do we have altogether?

0 1 2 ③ 4 5 6 7 8 9 10 11 12 13 14 15 ⑯ 17 18 ⑲ 20

We can start at 3 and count on 16.
Or we can start at 16 and count on 3.
We can add the numbers in any order.
The answer will always be the same.
It is quicker to count on from the larger number.

141

Practise

1 Use cubes to add the numbers in 2 different ways.

a 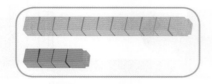 is the same as

11 + 4 =

b 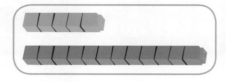 is the same as

5 + 12 =

2 Count on the number lines to add in 2 different ways.

a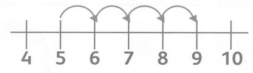

5 + 4 = ☐   4 + 5 = ☐

b

7 + 5 = ☐   5 + 7 = ☐

c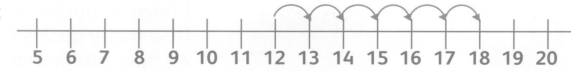

12 + 6 = ☐   6 + 12 = ☐

3 Solve these number stories.

a Gran buys 4 apples and 11 oranges.
How many fruits does she buy in total?

b Mr Khan buys 6 red pencils and 12 blue pencils.
How many pencils does he buy altogether?

Try this

Try Jack's ideas for these additions.

7 + 4 = ☐   3 + 11 = ☐

6 + 6 = ☐   12 + 5 = ☐   6 + 8 = ☐

> I always count on from the first number when I add.

What will you say to help Jack **improve** his method?

Subtracting on a number line

Explore

Count back 5 from different balloons.
What numbers do you land on?
Can you start on numbers that make your answer **more than** 10?
Can you start on numbers that make your answer **less than** 10?

Learn

Say the first number.
Then jump backwards as you **take away** the second number.

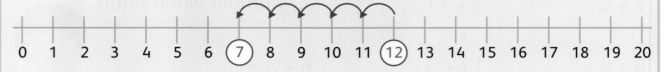

12 – 5 = 7
12 count back 5 is 7.
12 take away 5 is 7.

> **Maths word**
> take away

Practise

1 Count back on the number line to work out the answers.

0 1 2 3 4 5 6 7 8 9 10 11 12 13 14 15 16 17 18 19 20

a 11 – 4 = ☐

b 12 – 4 = ☐

c 13 – 4 = ☐

d 14 – 4 = ☐

2 David uses the starting numbers. He spins a spinner and counts back that number. What number does he land on each time?

	a	b	c	d
Starting number	15	19	14	16
Spinner number	3	6	4	5
Lands on				

Practise *(continued)*

3 Find the missing numbers. Use the number line to help you.

0 1 2 3 4 5 6 7 8 9 10 11 12 13 14 15 16 17 18 19 20

15 − [4] = 11

15 take away something leaves 11.
So, count back from 15 until you get to 11.

a 16 − [ ] = 13   b 17 − [ ] = 13

c 14 − [ ] = 9   d 13 − [ ] = 7

Let's talk

I know that 15 − 5 = 10,
so 15 − 6 will be less than 10.

I know that 16 − 6 = 10, so
16 − 5 will be more than 10.

Work with a partner.
Use Jack and Viti's thinking to estimate these subtractions.

18 − 6 = [ ]   19 − 7 = [ ]

14 − 5 = [ ]   17 − 9 = [ ]

Counting up on a number line to find the difference

Explore

What is the same and what is different in these pictures?

Learn

There are 11 yellow counters and 7 blue counters.

Both rows have 7 counters. The top row has 4 more counters.

7 counters   4 more counters

The difference between the two rows is 4 counters.
The difference between 11 and 7 is 4.
11 − 7 = 4
We can also show the difference by counting on from the smaller number on a number line.

0 1 2 3 4 5 6 ⑦ 8 9 10 ⑪ 12 13 14 15 16 17 18 19 20

Practise

1 Find the difference between the two rows of counters.
Write the matching subtraction sentences.

a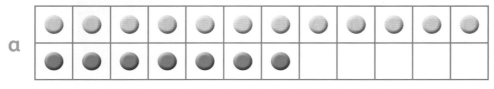

b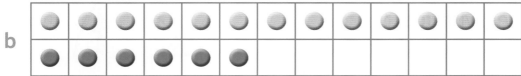

Practise *(continued)*

2   Find the difference between the pairs of numbers.
    Count up from the smaller number on the number line.

0 1 2 3 4 5 6 7 8 9 10 11 12 13 14 15 16 17 18 19 20

a 12 and 8   b 11 and 17   c 14 and 9   d 18 and 13

3   Solve these subtractions by finding the difference.

a 14 − 11 = ☐   b 15 − 12 = ☐   c 16 − 13 = ☐

d 17 − 11 = ☐   e 18 − 12 = ☐   f 19 − 13 = ☐

Try this

The difference between 2 numbers is 5.
What could the 2 numbers be?
Find at least 3 different solutions.

?   ?

Let's talk

I think it is helpful to find the
difference for subtractions with
numbers that are close together.

Do you agree with Jack?
Try out some ideas to help you decide.

Adding amounts up to 20

Explore

What different additions can you make?

Which additions will have answers **larger than** 10?

Which additions will have answers **smaller than** 10?

Learn

There are 8 red counters and 6 blue counters.

How many counters in total?

First we fill up a ten frame. Then we can see how many more than 10 we have.

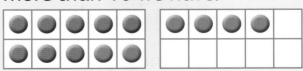

We have 10 and 4 more.

So, 8 + 6 = 14.

Practise

1 Use counters to find the totals. **Hint:** Fill up a ten frame first.

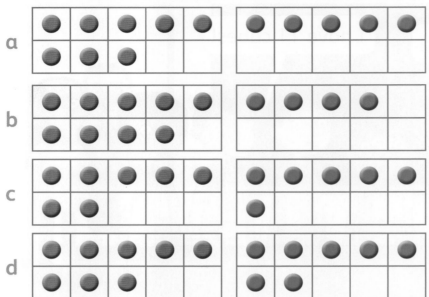

a

b

c

d

2 Use counters to complete these sets of additions.

a	b	c
7 + 4	8 + 5	9 + 4
7 + 6	8 + 6	5 + 9
7 + 8	8 + 7	6 + 9

Remember, you can put the larger number first to help you add.

3 Zara wants to plant seeds. She looks for flower pots.
She finds 4 under a bench and 7 next to a tree.
How many flower pots does she find in total?

Try this

Maris is adding 7 counters
and 5 counters.
What mistake has
Maris made?
How can you help her to correct it?

I need 4 counters to fill a ten frame and I have 1 counter left. I think 7 + 5 = 11.

Learn

We can use addition facts in another way to help us to add.
Look at the addition 12 + 5.

We can show this using cubes.

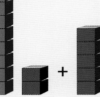

There is one 10 and some more.
2 + 5 = 7 is an addition fact we know.
So there are 10 and 7 more.
12 + 5 = 17

Practise

1 Use cubes to show each addition.
Remember to use addition facts to help you.

a 11 + 4　　　　b 12 + 6　　　　c 13 + 5

d 6 + 11　　　　e 2 + 17　　　　f 4 + 14

2 Find the missing numbers.

a 12 + ⬜ = 17　　b 11 + ⬜ = 16　　c ⬜ + 4 = 17

d ⬜ + 13 = 18　　e ⬜ + ⬜ = 15　f ⬜ + ⬜ = 18

3 David has 16 marbles.
More than 10 marbles are in a bag.
The rest are in his pocket.
How many marbles can be in the bag and in his pocket?
Write the addition facts that you used.

Try this

Jack uses the addition fact 6 and 3 is 9 to help him add.
Which of these additions does he do? How do you know?

(16 + 3)　(14 + 3)　(13 + 6)　(6 + 14)　(3 + 16)　(6 + 13)

151

Subtracting amounts up to 20

Explore

Maris, Viti and Zara are working in the school garden.

What subtractions do you see?

Maths word
subtract

Learn

The counters show 11 lettuces.

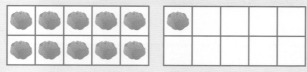

Rabbits eat 3 of the lettuces.

We can **subtract** 3 counters to show this.

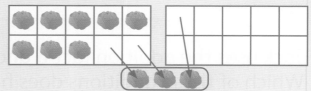

There are 8 lettuces left.
11 − 3 = 8

152

Practise

1 Show how you can answer these.

a  subtract

b  subtract

c subtract

2 Use counters to complete these subtractions.

15 − 5 = ☐   15 − 6 = ☐   15 − 7 = ☐

3 Jack and David are throwing balls into a box.

a David has 16 throws. He misses with 5 balls.
How many balls go into the box?

b Jack also has 16 throws. He gets 12 balls into the box.
How many does he miss?

Try this

Answer, then sort these subtractions in a table like the one below.

16 − 4 = ☐   17 − 9 = ☐   12 − 2 = ☐   15 − 5 = ☐

Answer is less than 10	Answer is 10	Answer is more than 10

Let's talk

Share your sorting for the Try this activity with a partner.
Look carefully at the subtractions in each box.
What do you notice about the ones **digit** in each number?
What patterns do you see? **Digits** are: 0, 1, 2, 3, 4, 5, 6, 7, 8, 9.

Doubling numbers up to 10

What do you notice about the groups of different animals in this picture?

How many seals?
How many birds?
How many penguins?

Maths words

twice
double

Learn

Doubling a number is the same as adding the same number **twice** or adding **2 lots of** the same number. For example: **6 = 3 + 3**

There are 3 birds.

3 more birds arrive.
There are 6 birds altogether.

You can show doubling on a number track. Start at 3 and count on another 3.

Double 3 is 6.

What is double 2?
Start at 2 and count on another 2.
Double 2 is 4.

Practise

1 Make hops on the number track to show double the number.

1	2	3	4	5	6	7	8	9	10

   a Double 2   **b** Double 3   **c** Double 4   **d** Double 5

2 Draw double of these shapes.

  a    **b**

3 Write each doubles number sentence.

  a ●●● + ●●●   6 = ☐ + ☐

  b ●● + ●●   4 = ☐ + ☐

  c  +   10 = ☐ + ☐

4 There are 4 toy trains.
There are double the number of toy cars.
How many toy cars are there?

Try this

Zara has double the amount of money that Viti has.
Zara has 8 cents. How much does Viti have?

Doubling numbers up to double 10

Explore

Look at all the things in the toy village!

Let's make a village with double the number of things in it!

What will be in Maris and Jack's new toy village?

Learn

How many paint tins?

If we **double** the number,
how many paint tins now?

There are now **twice** as many.
5 doubled is 10. 1 doubled is 2.
So, 6 doubled is 12 because:
10 + 2 = 12

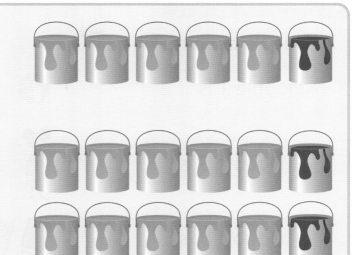

Practise

1 Double the number of paint tins each time.

a

b

c

2 Complete the sentences.

a There are ☐ cubes in the first chain.

b There are ☐ cubes in the second chain.

c Double ☐ is ☐

Practise *(continued)*

3   Double each number.

Double 1 is   2

a   Double 5 is   ☐

b   Double 8 is   ☐

c   Double 10 is   ☐

Try this

When you hold up 2 fingers in front of a mirror, you will see
4 fingers altogether.
How many fingers when you hold up 3 fingers?
How many fingers when you hold up 7 fingers?
What is the largest number you can see when you hold your fingers
in front of a mirror?

Let's talk

If you double a
number, you will
always get an
even number.

Do you agree with David?
Try some examples with a partner.
What did you find out?

Quiz

1 Find the missing numbers. What pattern do you see?

$4 + 1 = \boxed{}$ $4 + \boxed{} = 6$ $4 + 3 = \boxed{}$ $4 + \boxed{} = 8$

2 Make 10 on each side of the scales.

a 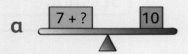   b

c 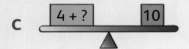   d

3 a Take a handful each of red and blue cubes. Make 2 towers. How many cubes in each tower? How many altogether?

b Check your addition. Change the order of your 2 towers.

c Add again. Did you get the same answer?

4 a Annay has 17 cubes. He takes away 8. Use the number line.

0 1 2 3 4 5 6 7 8 9 10 11 12 13 14 15 16 ⑰ 18 19 20

b Write the subtraction number sentence in your notebook.

$17 - \boxed{} = \boxed{}$

5 Take a card from a pack of 1 to 5 number cards.
Make a tower of cubes that is double the number on your card.
How tall is your tower?

      4 5

15

Shapes, direction and movement

Naming and sorting shapes

Explore

Maths words
2D shape
3D shape

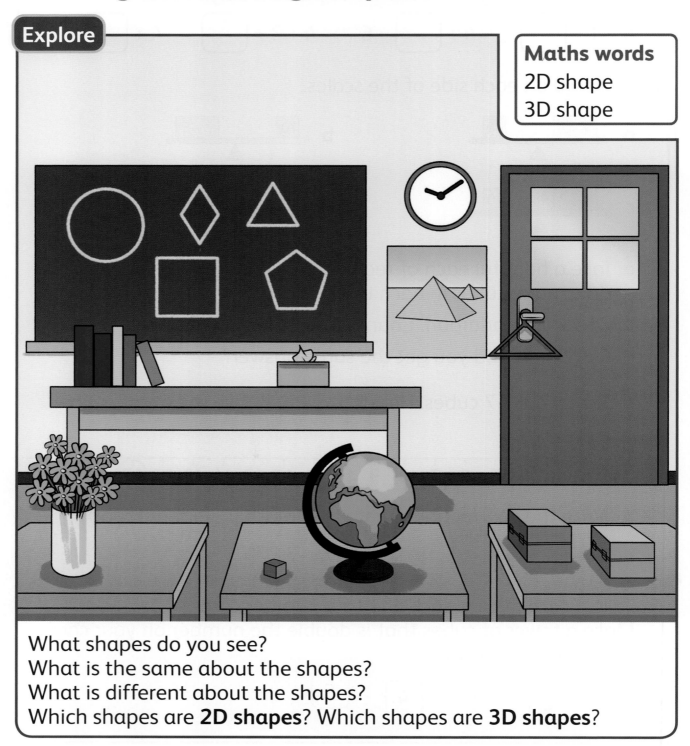

What shapes do you see?
What is the same about the shapes?
What is different about the shapes?
Which shapes are **2D shapes**? Which shapes are **3D shapes**?

Learn

2D shapes

Maths word
prism

circle 1 curved side

triangle 3 straight sides, 3 corners

square 4 straight sides of equal length, 4 corners

rectangle 4 straight sides, 4 corners

pentagon 5 straight sides , 5 corners

hexagon 6 straight sides , 6 corners

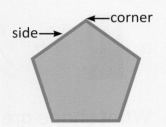

3D shapes

sphere 1 curved surface

cube 6 faces, 12 edges, 8 corners

cuboid 6 faces, 12 edges, 8 corners

cylinder 2 faces, 1 curved surface

cone 1 face, 1 curved surface, 1 corner

triangular **prism**  5 faces, 9 edges, 6 corners

square-based pyramid

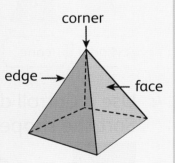

5 faces, 8 edges, 5 corners

Practise

1 **What shape is each object below?**

(circle) (triangle) (rectangle) (square) (pentagon) (hexagon)

a    b 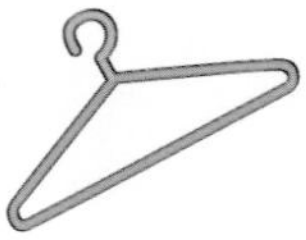   c    d    e 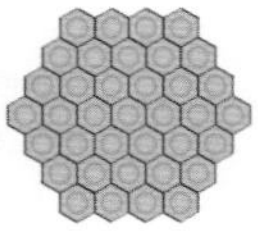   f

2 **What shape are these objects?**

(sphere) (cube) (cuboid) (cone) (cylinder)

a 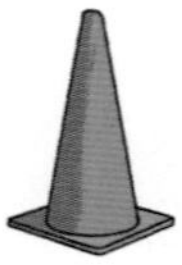   b    c 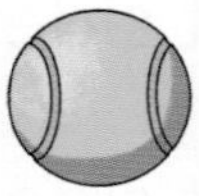   d    e

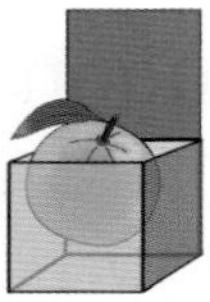

3 Use a Venn diagram like this.
Sort the shapes and add a heading.

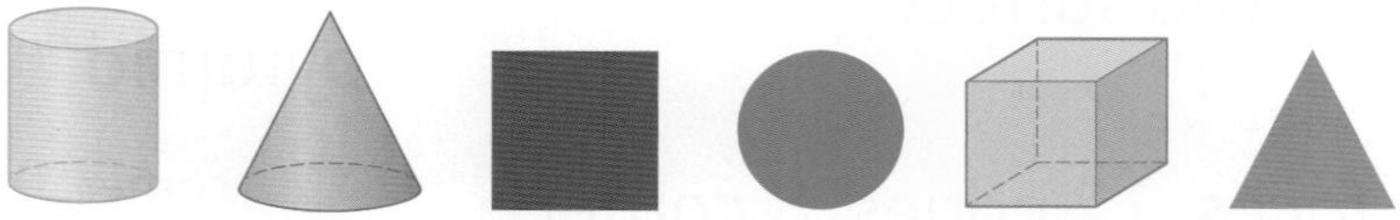

(cylinder) (cone) (square) (circle) (cube) (triangle)

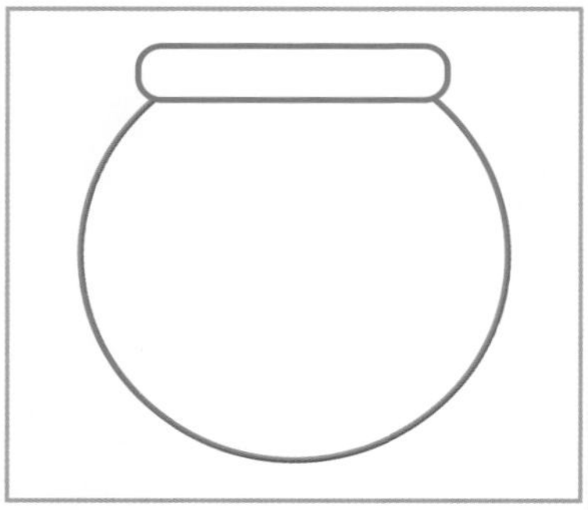

4 Use a Carroll diagram like this.
Sort the shapes and add headings.

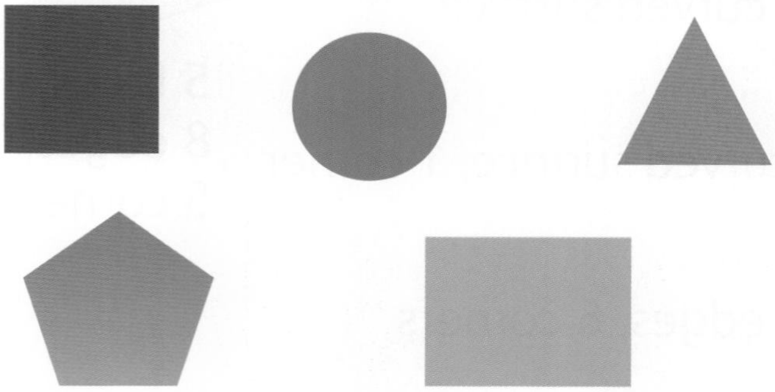

Practise *(continued)*

5 What numbers are missing in this table?

Shape	Number of sides	Number of corners
Triangle		3
Rectangle	4	
Pentagon		5

Try this

What shape is this?
What other 3D shapes can you make?
How many straws do you need to
make a square-based pyramid?

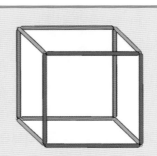

Let's talk

Choose a shape.
Describe the shape to a partner.
Ask your partner to guess the shape.

My shape has 6
sides and 6 corners.
What is my shape?

Sequences

Explore

David has made a **sequence**.
What ball is next in the sequence?
See if you can make a different sequence.

Maths word
sequence

Learn

We can make a sequence by repeating shapes in a pattern.

A twig is next in this sequence.

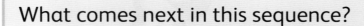

What comes next in this sequence?

A 5c coin is next in this sequence.

Practise

1 Draw the next shape in each sequence.

a
b

c
d

2 What is next each time?

a
b

c
d

Try this ⭐

Take a handful of counters.
Make a sequence.
Try to make a sequence using 2 different colours.
Try to make a sequence using 3 different colours.
Now use 3 colours to make a different sequence.

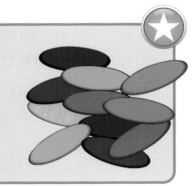

Let's talk ⭐

Jack says that the 8th shape in this sequence is a blue cube.
Is he correct?
Say why or why not.

Direction and movement

What shapes are the kites? What shape is the kite on the right? And on the left?
Do the shapes look the same when you **rotate** them or turn them around?

Maths word

rotate

Rotating shapes

To rotate means to turn.
When you turn a shape, it can look the same or different.

The square and the triangle look the same when you turn them.

The rectangle looks different when you turn it.

Directions

We can turn left or right.

left

right

We can go **backwards** and **forwards**.

backwards

forwards

Practise

1 Which shapes are the same when you rotate them?

a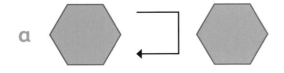

c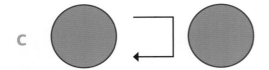

b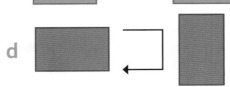

d

2 Start in square **B3** each time.
Follow the directions.
Which square does the duck walk to?

a 1 square forwards

b Turn left, 2 squares forwards

c 1 square backwards

d Turn right, 1 square forwards

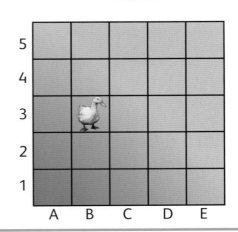

Try this

Put a counter in square **A1**.
Give your partner directions
to get to these squares.

a F5

b C3

c F6

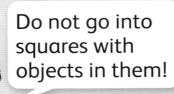

Do not go into
squares with
objects in them!

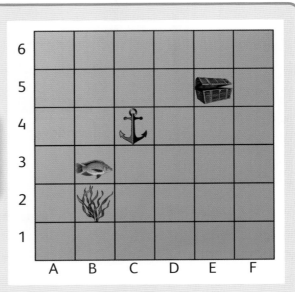

Quiz

1 Name each shape below.

hexagon   triangle   pentagon

square   cone   cube   pyramid

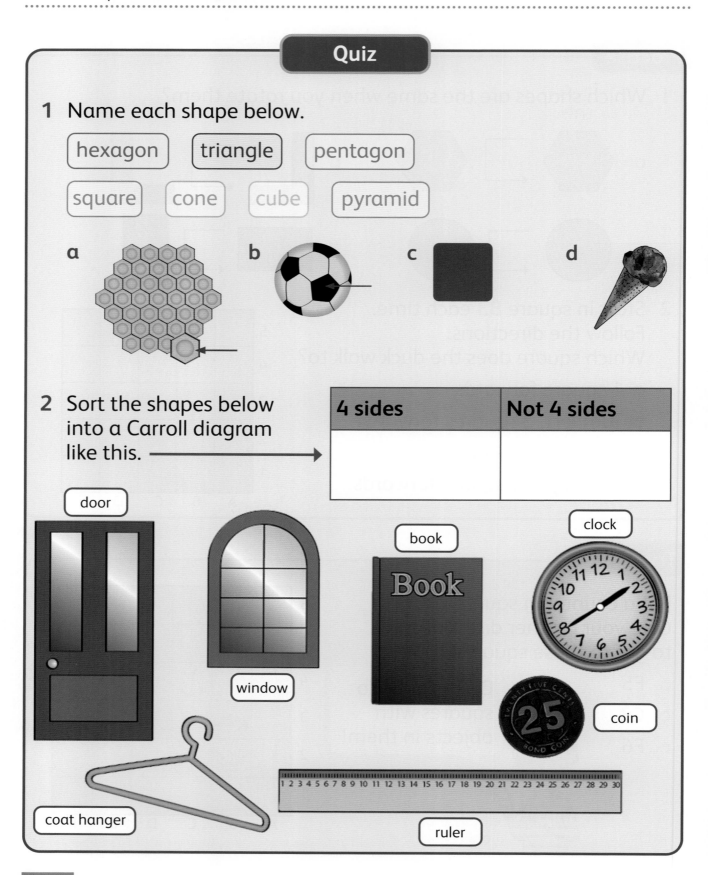

a

b

c

d

2 Sort the shapes below into a Carroll diagram like this. →

4 sides	Not 4 sides

door

window

book

clock

Book

coin

coat hanger

ruler

Quiz

3 Draw the next 2 shapes in each sequence.

a    b

c    d

4 Use these words to complete the sentences below.

| backwards | forwards | left | right |

a    Is the child walking forwards or backwards?

b    Is the child stepping forwards or backwards?

c    Is the cat on the right or the left?

16

Statistical methods

Pictograms, lists and tables

Explore

Annay and I travelled by bus.

How did you travel to school today?

How can we record this **data**?

Can we show the information in different ways?

How many children travelled by car?

Maths word
data

Learn

We can write a **list**.

Car: Guss, Elok, David
Bus: Maris, Annay, Jack, Jin
Bike: Zara, Viti
Walk: Sanchia, Pia

We can show the data in a **table**.

Transport	Number of children
Car	3
Bus	4
Bike	2
Walk	2

We can show the data as a **pictogram**.

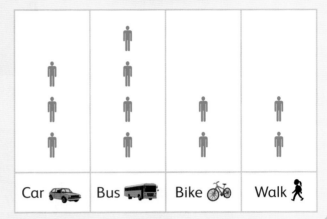

A pictogram to show how the children travelled to school

Car Bus Bike Walk

Practise

This pictogram shows how many cakes a baker sells each day.

Cakes a baker sells each day

1 How many did she sell on Thursday?
2 When did she sell the **most** cakes?
3 When did she sell the **fewest** cakes?
4 How many did she sell on Monday and Tuesday?
5 How many more cakes did she sell on Tuesday than on Friday?
6 How many cakes did she sell altogether?

Try this

How will you collect and record the data?

What is the favourite food in your class?

Draw a pictogram.

Maths words
list   table
pictogram

Venn diagrams and Carroll diagrams

What rule could you have for the Venn diagram?

How could you sort the numbers?

Venn diagrams

We can use **Venn diagrams** to sort numbers or objects. Where would the number 12 go in this Venn diagram? Can you explain why?

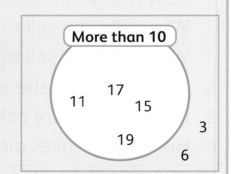

More than 10

Carroll diagrams

We can use **Carroll diagrams** to sort numbers or objects.

Odd numbers	Even numbers
3, 11, 15, 17, 19	6

Where would the number 8 go in the Carroll diagram above?

Practise

1 Sort the shapes into a Venn diagram like this.

a

b

c

d

e

f

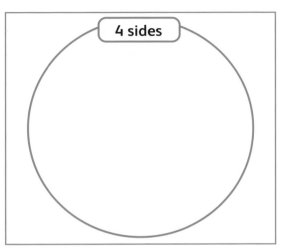

2

More than 9	Less than 9
12, 20, 16, 11, 10	3, 5, 4, 1, 8

Use a Carroll diagram to sort the numbers in a different way.

Try this

How many ways can you sort the numbers in a Venn diagram?

| 3 | 4 | 8 | 7 | 12 | 5 |

Let's talk

Talk about headings for this Carroll diagram.

6, 8, 4, 10	13, 19, 17, 15

Where would the number 9 go?

173

Block graphs

Maris, Viti and David are counting cubes.

How many blue cubes?

How many yellow and orange cubes altogether?

Whose cubes are easiest to count? Why?

Learn

Look at the **block graph.**
How many learners in total?
We can add the numbers
to find out.
7 + 6 + 3 + 8 + 2 = 26
There are 26 learners in total.

How many more learners like
yellow better than orange?
8 learners like yellow.
2 learners like orange.
8 − 2 = 6
6 more learners like yellow
better than orange.

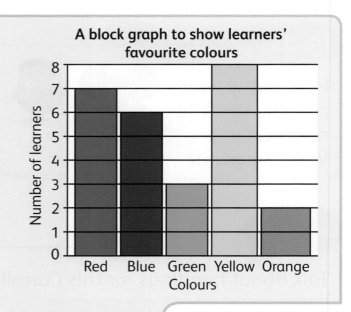

A block graph to show learners' favourite colours

Maths word
block graph

Practise

1 This block graph shows the favourite fruits of the Class 1 learners.

A block graph to show the favourite fruits of Class 1 learners

Fruits

(vertical axis: Number of learners, 0 to 7)

   a What is the favourite fruit in Class 1?

   b What is the least favourite fruit?

   c How many learners like bananas?

   d How many learners like grapes and bananas?

   e How many more learners like apples better than pineapples?

   f Which 2 fruits do learners like equally?

   g How many learners in total?

2 Draw a block graph to show the data in this table.

Favourite animal	Number of learners
Cat	6
Dog	3
Fish	4
Bird	2

Let's talk

Look at the data in **Practise** question 2.

What questions could you ask?

Quiz

1   The table and pictogram show how many books were sold.

Days of the week	Books sold
Monday	4
Tuesday	6
Wednesday	10
Thursday	2
Friday	7

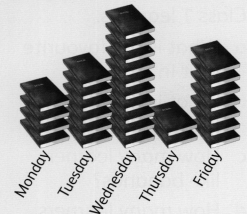

a   How many books were sold on Friday?

b   How many books were sold on Monday and Tuesday?

c   On which day were the most books sold?

d   On which day were the least books sold?

e   How many books were sold altogether?

f   How many more books were sold on Wednesday than
on Thursday?

2   Use a Venn diagram like this.

a   Write 2 numbers that go inside the hoop.

b   Write 2 numbers that go outside
the hoop.

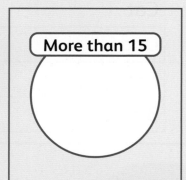

More than 15

3

2, 0, 4	7, 9, 11

a   What headings could you use for this Carroll diagram?

b   Choose 2 numbers. Add them to the Carroll diagram.

Halving numbers

Explore

Equal parts

Each side is $\frac{1}{2}$.

$\frac{1}{2}$ means **half**.

Add 1 counter on each fold.
How many in each **part**?
How many in total?
Add 1 more to each side. Repeat.

Maths words

half	part
share	equal
halve	

Learn

We can **share** 12 shells to make 2 **equal** groups.
We say we **halve** the group. Half of 12 is 6.

Can you share these 15 shells into 2 groups? Why or why not?

Practise

1 Trace the dotted lines with your finger.
   Then write $\frac{1}{2}$.

   $\frac{1}{2}$

2 What is half of each number of dots?

   a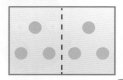

   Half of 6 = ☐

   b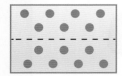

   Half of 14 = ☐

   c

   $\frac{1}{2}$ of 10 = ☐

   d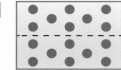

   $\frac{1}{2}$ of 16 = ☐

3 Complete these halving facts.
   Use counters to check.

   a

   b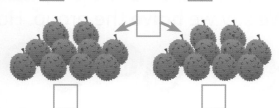

   c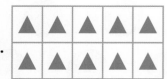
   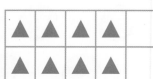

4 What is $\frac{1}{2}$ of 18?

   Use cubes or counters to check.

Understanding half and whole

Explore

Point to halfway between Maris and David.

Maths word
whole

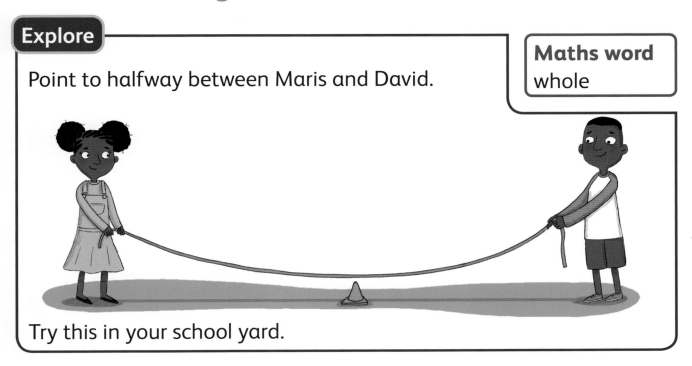

Try this in your school yard.

Learn

2 halves make a **whole**.

Practise

1 Fold a piece of wool, string or a skipping rope in half.
Try it with different lengths.

Practise (continued)

2 Find some 2D shapes and draw around them.
Draw a line on each shape and colour it in halves.

3 Draw a picture of 2 halves that are
joined to make a whole.

4 a Use 5 cubes to make a shape.
  b Use 5 more cubes to copy your shape.
  c Join both shapes to make a whole shape.

Try this

Which numbers can you share into halves?
Which numbers can you not share
into halves?

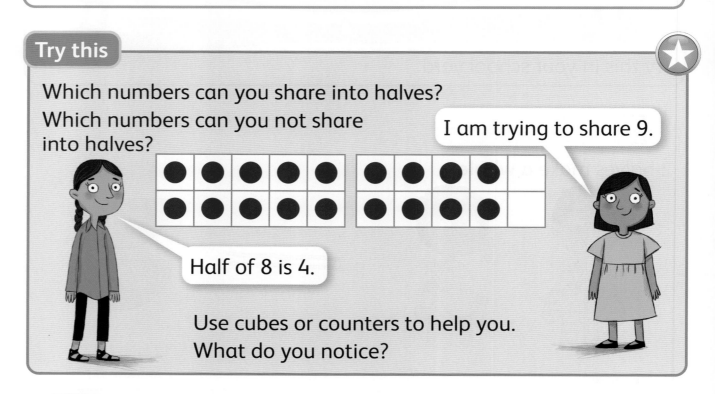

I am trying to share 9.

Half of 8 is 4.

Use cubes or counters to help you.
What do you notice?

Let's talk

Work with a partner. Look around your classroom or the school
yard for shapes or groups that are in 2 halves.

Quiz

1 What halving fact does this show?

2 Arrange these cards to make sense.
Write your number sentence.

$\frac{1}{2}$ 5   10

3 Take 20 cubes.
Make 2 equal towers.
How many cubes are in each tower?

4 Which halves go together to make 3 whole shapes?

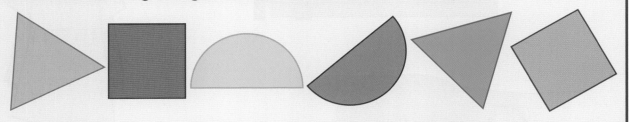

Time and measurement

Time

Explore

Where have you seen clocks or a watch like these?

Learn

We can use a clock or watch to tell the time.

There are 2 hands on a clock.
The big hand points to the **minutes**.
The small hand points to the **hours**.

minute hand

hour hand

When the minute hand
points to the 12,
we say the time is **o'clock**.
This clock shows that
the time is 3 o'clock.

When the minute hand
points to the 6,
we say it is **half-past**.
This clock shows that the
time is half-past 8.

Practise

What time does each clock show?

1   2   3   4

5   6   7   8

Try this

What time could it be in each picture?

a   b   c

SCHOOL

Let's talk

What time is lunchtime?

What time does the school day finish?

Capacity

Explore

Maths word
capacity

Learn

We can compare how much water a container holds.

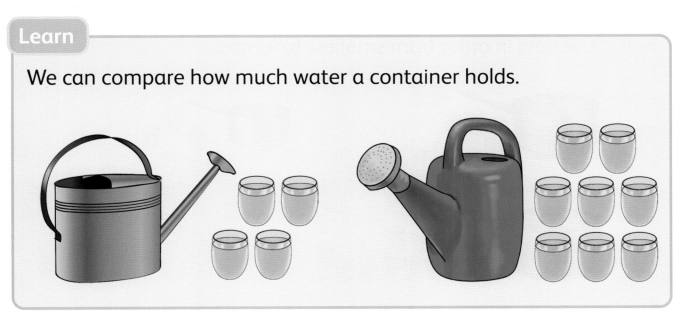

Practise

1 Which container holds more?

a

b

2 Put the jugs in order from smallest to largest.

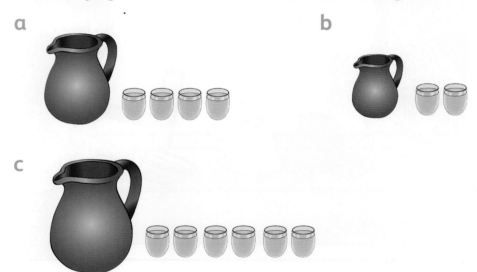

3 Put the pots in order from smallest to largest.

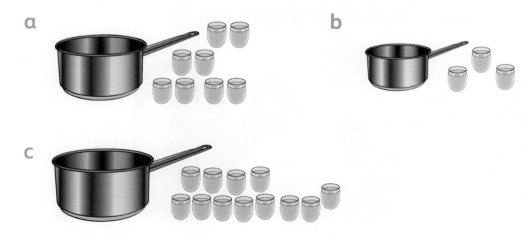

Try this

Pour cups of water into 3 different containers.

a Which container holds **more** than the others?
How do you know?

b Which container holds **less** than the others?
How do you know?

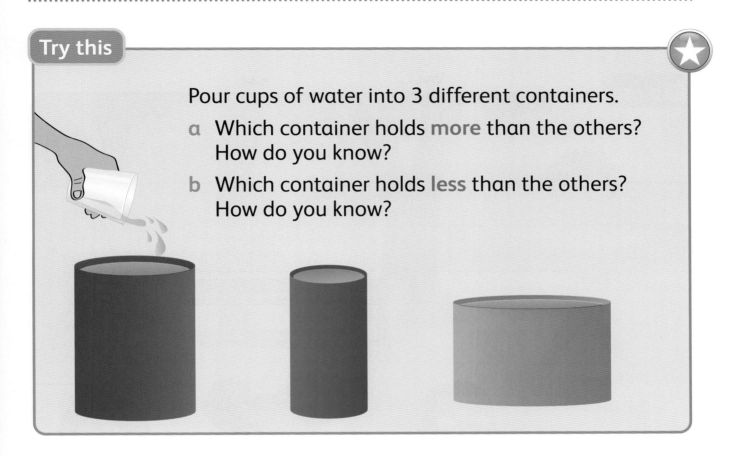

Let's talk

Jack is comparing the **capacity** of 2 plastic boxes.

What mistake has Jack made?

How can you help him to measure the capacity correctly?

The blue box holds 3 jugs of water. The yellow box holds 12 glasses of water. So, the yellow box holds more.

Measuring

Explore

What could you use to measure different things in this garden?

Learn

My parcel is 3 hand spans long.

My parcel is 20 cubes long. We cannot compare the lengths because cubes and hand spans are different units of measurement.

We can use these words to compare measurements.

longer

shorter

heavier   lighter

more   less

Practise

1 Which object is shorter?

2 Which object is lighter?

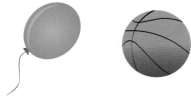

3 Which measuring instrument would you use to measure:

   a how hot it is today?

   b how heavy a parcel is?

   c how long a book is?

   d the capacity of a watering can?

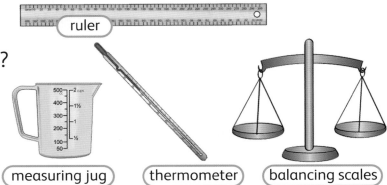

ruler

measuring jug   thermometer   balancing scales

Try this

David is comparing the length of the parcels. What is wrong? Why?

Let's talk

Talk to a partner.

a What could you use to measure the length of your classroom?

b What could you use to measure the capacity of a sink?

Quiz

1 Read the time on each clock.

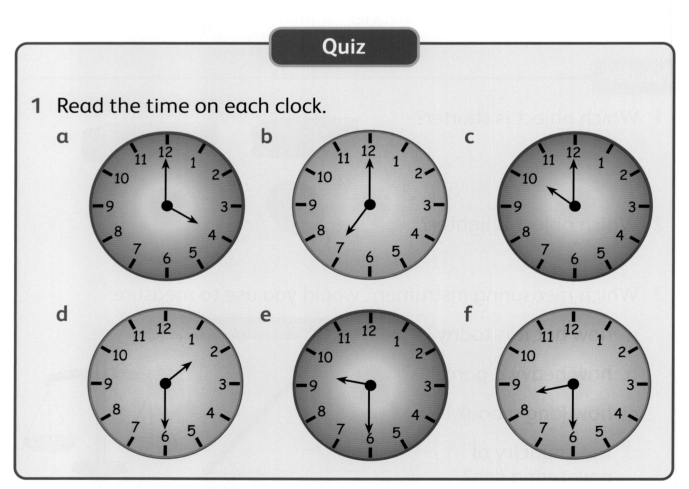

a   b   c

d   e   f

Quiz

2 Which container holds more?

 a

b

3 Put the containers in order from smallest to largest.

a

b

c

4 Which measuring instrument would you use to measure:

a the length of a shoe?

b how warm some water is?

c how much water is in a vase?

d how heavy a box is?

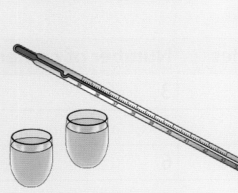 

Units 13–18

1 Put 2 cubes in a pot at a time. Count as you go.
When you have finished, count all the cubes to check.

2 Answer these additions and subtractions.

a 15 + 4 = ☐   b 17 − 4 = ☐

c 14 + 5 = ☐   d 17 − 6 = ☐

e 12 + 3 = ☐   f 15 − 4 = ☐

g 4 + 12 = ☐   h 13 − 5 = ☐

3 Copy and complete. The first one has been done for you.

Number	Double
5	10
4	
7	
	20

4 Copy and complete.

Shape	Number of sides	Number of corners
Triangle		3
Square	4	
Hexagon		6

5 a On which day were the most books sold?

b On which day were the least books sold?

c How many books were sold on Thursday?

d How many books were sold on Monday and Wednesday altogether?

e How many books were sold on Thursday and Friday?

f Write a question about the block graph.

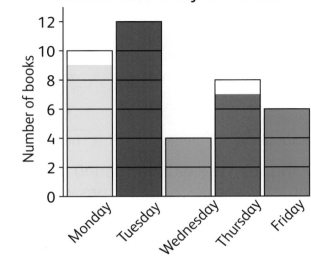

A block graph to show the number of books sold over 5 days of the week

Number of books

Days of the week

6 Take 19 counters.
Share them into 2 equal groups.
What do you notice?
Can you make half of 19? Why or why not?

7 What time does each clock show?

a

b

c

A half-past 4   **B** half-past 11   **C** 7 o'clock

2D shapes (two-dimensional) geometric shapes; flat shapes with sides and angles

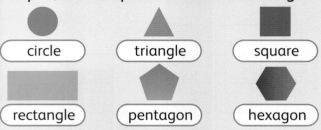

circle   triangle   square

rectangle   pentagon   hexagon

3D shapes (three-dimensional) geometric shapes; solid shapes with faces, edges and corners; see also *face*, *edge* and *corner*

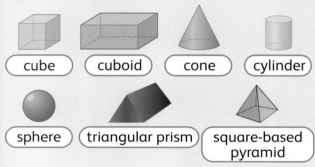

cube   cuboid   cone   cylinder

sphere   triangular prism   square-based pyramid

A

above on top of; higher than

add combine 2 or more numbers or objects to find a total

after behind, later

afternoon the time from noon or the middle of the day until the evening; see also *morning* and *evening*

altogether a collection of numbers or objects; in total

amount the number of objects or numbers

B

backwards moving in the direction behind you

before in front of, earlier

behind hidden by or further back than someone or something that is in front; see also *in front*

below under or at a lower level than

between in the middle of 2 numbers or objects

block graph a diagram that shows information

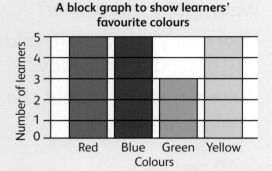

A block graph to show learners' favourite colours

Number of learners

Red   Blue   Green   Yellow

Colours

break up split or separate into pieces; or, change a sequence

C

capacity the amount a container holds

Carroll diagram a table used for grouping things

4 corners	Not 4 corners

cent(s) a coin value; see also *money*

circle a 2D shape with 1 curved side and no straight sides

coin a piece of metal used as money

combine put sets of things together so that they are a single set

compare to look at something to find similarities and differences

cone a 3D shape with a flat, circular face and a curved face; it has 1 corner

corner the point on a 2D shape where 2 sides meet

count give a number to each item in a set to find the total

count down start from the highest number and count backwards, for example: 10, 9, 8, 7, 6, 5, 4, 3, 2, 1

count up start from the lowest number and count forwards, for example: 1, 2, 3, 4, 5, and so on

cube a 3D shape made from 6 square faces

cuboid a 3D shape made from 6 rectangles, such as a cereal box; some of the rectangles could be squares; see also *cube*

curved a line that is not straight, such as a circle, or a surface that is not flat, such as an egg

cylinder a 3D shape with circular ends and 1 curved face joining the 2 circular ends

D

daily routine things that you do every day, such as after getting up, before going to bed, before and after eating or preparing food

data the pieces of information collected, for example, after asking a group of friends what their favourite food is

day 24 hours, starting from 12 o'clock midnight; see also *o'clock*

difference how much bigger or smaller one quantity is compared to another; usually found by subtraction, for example: the difference between 7 and 5 is 2; 7 − 5 = 2

digit the symbols 0, 1, 2, 3, 4, 5, 6, 7, 8, 9

double 2 lots of something, multiply by 2; twice as many

down from a higher to a lower point or number

E

edge the line made where 2 faces of a 3D shape meet; see also *face*

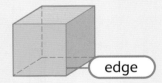

edge

equal the same in size or amount

equal part a piece that is the same size as something else; see also *part*

equals symbol: (=) means to have the same value as, for example: 5 + 3 = 7 + 1

estimate a sensible guess at how many or how much

even a whole number that can be grouped in twos; it is a multiple of 2; all numbers ending in 0, 2, 4, 6 or 8; see also *odd*

evening the period of time at the end of the day from about 6 o'clock (after the afternoon); see also *afternoon*

F

face a flat surface on a 3D shape; see also *edge*

fifth the number five (5) in order

first before anything or anyone else; the number one (1) in order

forwards moving in the direction you are facing, straight in front of you

fourth the number four (4) in order

fraction a part or amount of something; not a whole number, for example: $\frac{1}{2}$; see also *part*

H

half when a whole is divided into two equal parts

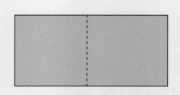 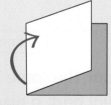

half-past a word to describe time, for example, half-past 10 is 10:30 or 30 minutes past the hour

halve cut or divide to make half

heavy, heavier, heaviest words used when describing mass

height see under *length*

hexagon a 2D shape with 6 straight sides

hour symbol: (h) A measure of time (60 minutes); see also *minute* and *second*

hour hand the short hand on a clock that measures the hours; 1 complete turn takes 12 hours; see also *minute hand*

I

in front a position that is just ahead of or before someone or something

L

larger bigger than

left the direction in which this arrow points

left over when something is remaining, after eating a meal or dividing a number

length, height words used to describe how long or tall something is; how far from 1 point to the other

less a smaller amount

less than used when comparing the size of 2 numbers or things, for example, 7 is less than 10; see also *more than*

lighter an object that has less mass than another object

lightest something that has the lowest mass of 3 or more things

list names or things that are often written one under the other

long measuring a great distance from 1 end to the other; see also *length, height*

long, longer, longest words that describe the length or height of different objects

M

mass the measure of the amount of matter in an object; usually measured in grams (g) or kilograms (kg)

measure to find out the size of something

minute symbol: (min.) a measure of time, there are 60 minutes in 1 hour; see also *second* and *hour*

minute hand the long hand on a clock face that measures the minutes; see also *hour hand*

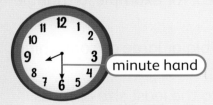

minute hand

money coins and notes used to buy things with; see also *cent(s)*

month there are 12 months in a year: January, February, March, April, May, June, July, August, September, October, November, December

more a bigger or larger amount

more than used when comparing the size or total of numbers or things; 10 is more than 7; see also *less than*

morning the time of day between midnight and noon or midday; see also *afternoon* and *evening*

N

night the time from sunset to sunrise in every 24-hour day; see also *day*

noon 12 o'clock in the day; midday; see also *morning*, *afternoon* and *evening*

note(s) a piece or pieces of paper used as money

number there are many different types of numbers, including counting numbers 0, 1, 2, 3 and so on; ordinal numbers 1st, 2nd, 3rd, and so on

number sentence a sentence of numbers and symbols, for example: $6 - 3 = 3$

O

o'clock a way of describing an hour time, for example, 5 o'clock; the minute hand always points to the 12; see also *half-past*, *hour hand* and *minute hand*

3 o'clock

odd all numbers ending in 1, 3, 5, 7 or 9

order put things in their correct place, following a rule

organise arrange or sort in an orderly way, to make things easy to find or do

P

pair 2 of something

part a piece of something; see also *whole*

pattern numbers, shapes or symbols that are repeated and follow a rule

pentagon a 2D shape with 5 straight sides

pictogram a type of chart where pictures are used to stand for quantities

Cakes a baker sells each day

Monday Tuesday Wednesday Thursday Friday

prism a solid shape with 2 identical ends and flat rectangular sides

pyramid a 3D shape with 5 faces and 5 corners; a square-based pyramid has a square base with 5 faces, 5 corners and 8 edges

Q

quick fast

R

rectangle a 2D shape with 4 straight sides; a square is a special type of rectangle with all 4 sides the same length

right the direction in which this arrow points

rotate turn or move in a circle

S

second a period of time; there are 60 seconds in 1 minute

second the number two (2) in order

sequence when objects or numbers are arranged in a particular order to form a pattern

share to put objects into equal-sized groups, 1 at a time

shorter not as long or tall as another object or person; see also *length, height*

shortest having the least length or height of 3 or more objects or people

side 2D shapes have sides, which can be straight or curved, for example, a triangle has 3 sides

smaller not as big or large as something else

smallest having the least size or mass when comparing 3 or more things

sort to put objects, shapes or numbers into different groups that follow the same rules

sphere a 3D shape like a ball

square a shape with 4 sides of equal length and equal angles

straight when something has no curves or corners; a straight line, for example, can be drawn using a ruler; see also *curved*

subtract to take away something from another thing

T

table a way to organise numbers or objects

take away another name for subtract; see also *subtract*

tall, taller, tallest words used when comparing heights, for example: Zara is tall but Viti is taller than she is, and Maris is the tallest

third the number three (3) in order

time how long something takes or lasts; time is measured in units such as seconds, minutes, hours, days, weeks, months and years

total the answer to an addition number sentence, for example, for 10 + 6, the total is 16

triangle a 2D shape with 3 straight sides

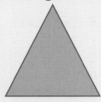

turn change direction, for example: change from facing forwards to facing left

twice two times

U

under directly below or less than

up towards a higher place or position; see also *count up*

V

value how much something is worth

Venn diagram a diagram with a circle and a rectangle around it to show sets

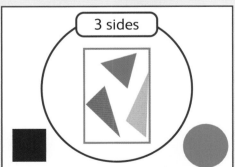

W

week there are 7 days in a week: Monday, Tuesday, Wednesday, Thursday, Friday, Saturday and Sunday

whole all of something; the entire thing

Y

year 365 (or 366) days make a year; the calendar shows all the months of a year

	JANUARY						
M	T	W	T	F	S	S
					1	2	3
4	5	6	7	8	9	10
11	12	13	14	15	16	17
18	19	20	21	22	23	24
25	26	27	28	29	30	31

	FEBRUARY						
M	T	W	T	F	S	S
1	2	3	4	5	6	7
8	9	10	11	12	13	14
15	16	17	18	19	20	21
22	23	24	25	26	27	28
29						

	MARCH						
M	T	W	T	F	S	S
1	2	3	4	5	6
7	8	9	10	11	12	13
14	15	16	17	18	19	20
21	22	23	24	25	26	27
28	29	30	31			

	APRIL						
M	T	W	T	F	S	S
				1	2	3
4	5	6	7	8	9	10
11	12	13	14	15	16	17
18	19	20	21	22	23	24
25	26	27	28	29	30	

	MAY						
M	T	W	T	F	S	S
						1
2	3	4	5	6	7	8
9	10	11	12	13	14	15
16	17	18	19	20	21	22
23/30	24/31	25	26	27	28	29

	JUNE						
M	T	W	T	F	S	S
	1	2	3	4	5
6	7	8	9	10	11	12
13	14	15	16	17	18	19
20	21	22	23	24	25	26
27	28	29	30			

	JULY						
M	T	W	T	F	S	S
				1	2	3
4	5	6	7	8	9	10
11	12	13	14	15	16	17
18	19	20	21	22	23	24
25	26	27	28	29	30	31

	AUGUST						
M	T	W	T	F	S	S
1	2	3	4	5	6	7
8	9	10	11	12	13	14
15	16	17	18	19	20	21
22	23	24	25	26	27	28
29	30	31				

	SEPTEMBER						
M	T	W	T	F	S	S
		1	2	3	4
5	6	7	8	9	10	11
12	13	14	15	16	17	18
19	20	21	22	23	24	25
26	27	28	29	30		

	OCTOBER						
M	T	W	T	F	S	S
				1	2
3	4	5	6	7	8	9
10	11	12	13	14	15	16
17	18	19	20	21	22	23
24/31	25	26	27	28	29	30

	NOVEMBER						
M	T	W	T	F	S	S
1	2	3	4	5	6
7	8	9	10	11	12	13
14	15	16	17	18	19	20
21	22	23	24	25	26	27
28	29	30				

	DECEMBER						
M	T	W	T	F	S	S
		1	2	3	4
5	6	7	8	9	10	11
12	13	14	15	16	17	18
19	20	21	22	23	24	25
26	27	28	29	30	31	

Z

zero nothing or 0

Thinking and working mathematically (TWM) skills vocabulary

characterising identifying and describing the mathematical properties of an object

classifying organising objects into groups according to their mathematical properties

conjecturing forming mathematical questions or ideas

convincing presenting evidence to justify or challenge a mathematical idea or solution

critiquing comparing and evaluating mathematical ideas, representations or solutions to identify advantages and disadvantages

generalising recognising an underlying pattern by identifying many examples that satisfy the same mathematical criteria

improving refining mathematical ideas or representations to develop a more effective approach or solution

specialising choosing an example and checking to see if it satisfies or does not satisfy specific mathematical criteria